Science and Uncertainty

British Library Cataloguing in Publication Data

Science and uncertainty.
1. Science——Philosophy
I. Nash, Sara
501 Q175

ISBN 0-905927-96-6

Published by Science Reviews Ltd.,
40 The Fairway, Northwood, Middlesex, England

Typeset by Geoffrey Morton MIOP
Brightwell, Wallingford, Oxfordshire

Printed by Henry Ling Ltd., The Dorset Press,
Dorchester, England.

Science and Uncertainty

Edited by Sara Nash

Proceedings of a Conference
held under the auspices of IBM United Kingdom Ltd.,
London, March 1984

IBM United Kingdom Ltd.

Science Reviews Ltd.
1985

Acknowledgements

We gratefully acknowledge permissions granted by the following to reproduce text or illustrative material: Springer Verlag, Inc. New York for permission for *Beginnings of Organic Evolution* from *The Study of Time* Vol. IV (eds J.T. Fraser, N. Lawrence and D. Park) New York 1981, in *On the Rocks: The Vital Cocktail* by G. Cairns-Smith; The Milne Society and The Royal Astronomical Society for *Statistics, Geometry and the Cosmos* (the Milne Lecture, in *Quarterly Journal of the Royal Astronomical Society, 25*, 147–156, 1984) and Professor S. Chandrasekhar, Mrs. J.H. Weston-Smith and Ramsey & Muspratt for photographs in *Chance or Design?* by D. Kendall; John Wiley & Sons Ltd. for one figure from G. Adam, *Consciousness and Self Regulation* Vol. 2 (eds G.E. Schwartz and D. Shapiro) Chichester 1978, in *Consciousness: the Illusion of Certainty* by N. Dixon; For *Man, Technology and Uncertainty* by R. Garwin from *The Arms Race at a Time of Decision* (eds Joseph Rotbler and Alesandro Pescolini) © Pugwash Conferences on Science and World Affairs 1984 and reprinted by permission of Macmillan, London and Basingstoke and of St Martin's Press, Inc. New York; The Institute of Mathematics and its Applications for *The Application of Fractal Ideas to Geophysical Phenomena* in *Bulletin of the Institute of Mathematics and its Applications*, Vol. 20 in the chapter of the same title by P.A. Burrough; North-Holland Publishing Co. Amsterdam for pages 711–717, *Signal Processing II: Theories and Applications* (ed. H.W. Schussler) © Eurosip 1983, in *Digital Processing and Computational Complexity* by H. Nussbaumer.

Contents

Leslie Banks

Leslie Banks is Support Manager in IBM UK Academic Systems Marketing, and has chaired the annual interdisciplinary conference series, from 1981.

He joined IBM in 1963, after service in Local Government, Royal Artillery, Royal Air Force, Royal Navy, and United States Air Force.

Occasional spells with the University of London, the first attempt at an academic life in chemistry being interrupted by World War II, followed much later by BSc mathematics qualification and post-graduate diploma in computing science and numerical analysis.

His various IBM posts have included work involving research and academic institutions.

DFC, AFC, Croix de Guerre
Fellow, Royal Institute of Navigation (invention of navigation techniques)
Member of RGS Expeditions Committee, under Lord Schackleton, 1961–62.
Member, Royal Institution, and Society for Psychical Research.
Publications: "Pressure Pattern Flying" RIN; "Polar Air Navigation Guide" RAF; "Grid in all Latitudes" USAF; "An Aspect of Information Theory" IBM.

Introduction

This is the fourth in an annual series of interdisciplinary conferences launched in 1981 with *Science and Imagination* – a side to every question. The series is supported by IBM and aims to bring together established practitioners from a variety of disciplines and background to meet, discuss and argue in an environment removed from the pressures of everyday.

As on previous occasions, the contributors and listeners at *Science and Uncertainty* represented a wide spectrum of universities, research institutions and industry, their disciplines ranging from pure mathematics to biochemistry.

Science and Imagination in 1981 was followed by *Science and the Unexpected* in 1982, and by *Science and Synergy* in 1983. 1982 explored the nature of scientific creativity, exemplified by the Chinese ideogram for "crisis" combining the symbols of danger and opportunity, while 1983 examined the quality of synergy in the context of scientific discovery.

Thus up to 1983 the conference tapestry showed the imaginative leap, the sudden, unexpected discovery which also permits disappointment,

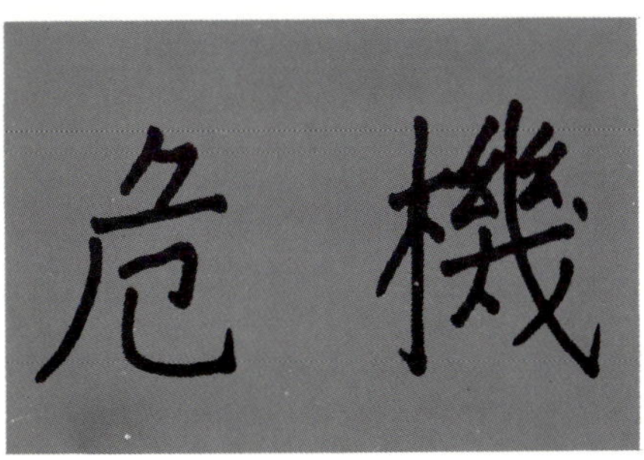

The Chinese ideograph combining *danger* and *opportunity* to signify *crisis*, typifies the best in scientific creativity.

Greek syn (together) and ergon (work) the parents of synergy.

Borromean Rings illustrate the loose connection between reality, the observer and the foundation of the observer's belief.

and in *Science and Synergy* we recognised the exponentiating yield resulting from the healthy convergence of ideas, disciplines and methodologies. But in science, as in everything else, all good things (and all bad things) eventually reach limits of applicability and we have seen that re-examination of fundamental concepts, paradigms, brings its rewards – and its pains – in those domains. We meet some spectacularly non-linear effects of intellectual excursions in uncertainty space, and indeed, in unpopularity space, arising from what Thomas Kuhn calls the incommensurability of rival paradigms.

The chosen symbol of *Science and Uncertainty*, the Borromean Rings, illustrates the loose connection between reality (whatever that may be) the observer (who often believes himself to be part of that reality) and the foundation of that belief: the observer's theory or paradigm.

If one ring is removed, the other two fall separately to the ground – or the ground comes up to collect them.

The theme of this conference suggests that all disciplines have essentially the same Borromean Ring structure. Indeed, we find the word *truth* appearing less and less in literature and seems to be replaced more and more by the word *model*. A model is a pattern; paradigm is a model. And the notion that we do not perceive reality directly is not new. Eddington said that it is not nature itself that we observe, but a nature transformed by our selective powers, a selection process which fits the processes of nature into a form or pattern largely of our own choosing.

But long before Eddington, in the eighth century, the founder of Tantric Yoga had already said that all appearances are self-contained in the mind, like reflections in a mirror.

This conference enquired into two special sub-themes: "Expectation and Refutations" and, analogously, "Noise and Signal". The analogy is suggested by Popper's thesis that the most meaningful signal is a refutation of a falsifiable hypothesis. Corroboration is not such a powerful signal.

Refutation of a *noisy* paradigm, no matter how well established and orthodox, is a positive step, a step which may require us to embrace a seemingly outrageous new paradigm. As Sherlock Holmes would say, we are left with the only non-impossible, *truth*. We need to renew our vows to keep an open, uncommitted mind – once a year at least.

Professor Eric R Laithwaite

Eric Laithwaite is Professor of Heavy Electrical Engineering at Imperial College of Science and Technology, London. Born in Atherton, Lancashire in 1921, he was educated at Kirkham Grammar School and the Regent Street Polytechnic. He joined the RAF in 1941, and from 1944 to 1945 worked on automatic pilots at the Royal Aircraft Establishment at Farnborough. After the war he graduated at Manchester University, being subsequently appointed Lecturer, then Senior Lecturer in Electrical Engineering, and in 1964 he took up his appointment at Imperial College. Amongst other appointments, he has held the posts of External Professor of Applied Electricity at the Royal Institution from 1967 to 1976, and President of the Association for Science Education. He is a director of Linear Motors Limited, Landspeed Limited, and Cotswold Research Limited, and consults for a large number of industrial organisations on an ad hoc basis.

In 1966 he was awarded the S.G. Brown Gold Medal of the Royal Society for inventions relating to linear induction motors. He has written over 270 papers and articles, ten books on electrical engineering and allied subjects, and is co-author of a book on his hobby – the butterflies and moths of the world. He frequently appears on television in such programmes as "Tomorrow's World", "Horizon", etc.

A scientist is often wrong but never in doubt

The history of science is littered with attempts to make it tidy. The time interval between all matter being said to be made up of earth, fire, air and water and its four fundamental constituents being declared to be protons, neutrons, electrons and positrons, occupies many centuries. But the degree of tidiness remains the same in that there seems to have been, until the 1930s, a deliberate attempt to limit the 'fundamentals' to *four*. The underlying feeling that, as Sir James Jeans once put it, "God is a mathematician" always seems to be there.

Scientists themselves built up a mystique having certain facets in common with the rituals of Black Magic no less, and one of these was called 'The Scientific Method'. It took a man of medicine, Sir Peter Medawar, to lay this ghost in a series of lectures in the US, which was subsequently published.[1] Of the Scientific Method, he had this to say:

> "Ask a scientist what he conceives the scientific method to be, and he will adopt an expression that is at once solemn and shifty-eyed; solemn, because he feels he ought to declare an opinion; shifty-eyed, because he is wondering how to conceal the fact that he has no opinion to declare."

Half a century earlier the great engineer Osborne Reynolds had expressed a similar sentiment in a lecture to the Society of Arts when he declared:[2]

> "I have to deal with facts, and I shall try to deal with nothing but facts. Many of these facts, or the conclusions to be immediately drawn from them, may appear to bear on the possibilities — or, rather, the impossibilities — of art. But in the Society of Arts I need not point out that art knows no limit; where one way is found to be closed, it is the function of art to find another. Science teaches us the results that will follow from a known condition of things; but there is always the unknown condition, the future effect of which no science can predict."

In this context, Reynolds clearly used the word 'art' to mean engineering, one might almost say, as *opposed* to science.

But the tidy structure of physics began to crumble, slowly at first, and then with increasing rapidity as each new generation of investigators demonstrated more and more ridiculous facts that appeared quite logical. Niels Bohr, attacking Pauli at a meeting in New York, on behalf of the younger scientists of the day began with the words: "We are all agreed that what you are saying is crazy. The thing that divides us is whether or not it is sufficiently crazy to have a chance of being correct." Subsequent events showed that it was *both* sufficiently crazy, *and* correct!

Oliver Heaviside, an acknowledged genius in his lifetime, had many of his early, highly mathematical papers declined, but as reported in the IEE Centenary Volume celebrating his birth, he was asking his mathematical colleagues to take on board such concepts as taking the square root of the process of partial differentiation. These days a good sixth form pupil, being told of this, might remark in a rather bored manner, "Oh, you mean $\sqrt{p}$." We do progress, after all.

Mass

Yet some of the old dogma remains almost inextricably mixed with the new thinking and is fed to each new generation of young people as if it really were the *truth*. Perhaps the very worst example is the usual school physics definition of mass:

"Mass is the amount of matter in a body."

Let us examine just what this means. The bright pupil is entitled to ask at once what the word 'amount' means. "Is it volume?" — of course not or that would have been stated. "Do you count the fundamental particles?" The answer is less obvious but after some thought the answer given will be, "Well, actually — NO!" So the word 'amount' has, apparently, no scientific meaning — according to the scientific method — before one even tries to tackle the even more difficult question, "What is matter?".

To answer that matter is anything that occupies space is begging the question for it only leads to the answer to "And what is space?" — "Space is to put matter in!".

Yet if mass is to be one of our three fundamental dimensions to take its place alongside 'length' and 'time', an answer as to what it is must be forthcoming. Quite apart from this reason a very young child soon gets the idea of subdividing matter (better described at that age perhaps as 'stuff') by cutting it in half, and then in halves again and so on. Soon the child is eyeing specks of dust and asking the question: "Are those the smallest bits of matter there are?".

We give the child a lens; at once it can see smaller specks of dust that it could not see without the lens. But the rot has set in. Are these new specks really there, or was it a magic property of the glass that gave an illusory picture? From now on the evidence for smaller 'bits' is going to become more and more circumstantial and speculative and less and less factual. However, the first hurdle can be surmounted without too much difficulty. Under the lens, a number of the otherwise invisible bits can be pushed together to form a larger bit, which can then be seen when the lens is removed.

But this becomes more difficult to do when the level of microscope is reached, and virtually impossible at electron microscope level where we approach the molecule still with the confidence that chemistry gives us. But the atom is definitely more difficult, for the pupil being taught the subject as if an atom of hydrogen were like an orange in London and a pinhead in Bristol orbiting the orange, has the right to ask: "What shape is the pinhead?" (i.e. the electron) and the only answer forthcoming will be, "An electron has no shape", for were it to be accorded a shape, the human mind will at once be aware of the possibility of subdividing it, which so far has proved impossible, even though it is now acknowledged that there are smaller bits than electrons! Incidentally, the 'oranges' are not solid but made up of 'neurons' stuck together with 'gluons'.

What then shall be said of 'quarks', of different kinds of quark, of 'charm', of 'up', of ——. As an engineer one feels the desire to scream: "Stop! It might not *be* like that!"

Let us examine the circumstantial evidence that led to these modern concepts. In the early 1930s, there were only protons, neutrons and electrons — *three* fundamentals only. Then, the positron was discovered and that really tidied things up, for now there was a '+' to balance the '−' and we were back to *four* fundamentals. Tidiness had prevailed.

But if we go back in time to when light was considered to be a stream of particles (the Corpuscular Theory) we find a prime example of the title of this paper. The particle theory could not explain the phenomena (facts) of interference and diffraction and the scientific world waited a long time before Thomas Young produced his wave theory which 'explained' all the old facts, plus interference and diffraction. After that it was all *waves*. It had to be tidy, of course. But as we now know, the baby was thrown out with the bathwater. Came others, Niels Bohr and Max Planck and it emerged that light was only sometimes to be regarded as waves and at other times as particles. The situation was described with great clarity by Sir Lawrence Bragg in some unpublished notes:

> "The tidy structure of classical physics has undergone a great change and reappraisal during the last fifty years. It is generally considered that the first crack appeared in its armour when J.J. Thomson discovered the electron. . . .

"The story was rounded off by the discovery of electron diffraction by G.P. Thomson, and Davisson and Germer. Electrons scattered by matter exhibit interference patterns of a form which would be accounted for by their being waves. Particles behave like waves, and waves like particles.

"The relation is a subtle one. When we postulate an experimental set-up, and wish to prophesy the result, we must treat both light and matter as waves. The nature of physical reality is such that we can only calculate the relative probability of an effect in various places. On the other hand, when we write the history of what did actually happen in an experiment, it is a history of particles whether of matter or light. A wavelike uncertain future, only expressible in probabilities, forever streaming through the moment 'Now', is transformed into a definite past of particles. Determinism takes on a new meaning."

But to return to our rather shaky model of a hydrogen atom with its one shapeless electron. Someone asked, "Would it be possible to extract the nuclei and pack them together with no space between them and, if so, what would be the density of such matter?" It emerged that a piece of such 'stuff', the size of a speck of dust, would have a mass greater than that of the Earth. "Can such matter exist?" asked the scientists. The astronomers got out their telescopes and began to search the Universe for it and came across — the neutron stars, with exactly the calculated density.

Now when an event like that happens, God is really in His Heaven and all's right with physics! — or was, until they made better telescopes and looked further, and found — the *quasars*. These objects had a density as great in comparison to a neutron star as a neutron star had in relation to ordinary matter on Earth. Panic set in: "My God, the oranges are not solid!" It was about this time that classical physics fell into ruins, even if quantum mechanics had not begun the process already.

Analogy

One analogy of the Scientific Method is that of an American who comes to England to discover the laws of cricket merely by watching the game and looking at the official score sheets on the following days. He must not converse with anyone on the subject, nor read accounts of the matches in newspapers, for in the scientific game this would be equivalent to asking God for the answers — and getting them.

The idea of being bowled out, of scoring runs and being caught out would be discovered on the first day, although caught behind the wicket, especially off the batting glove, would take considerably longer. He would realize that the fielders were not actually throwing the ball at the batsmen when they ran, but trying to hit the stumps! The rarer ways of being out such as 'hit wicket', 'handled ball' and the idea of declaring an innings closed might take several matches before he actually saw the phenomena. By the end of a season he is fairly confident that he knows 'nearly all of it' (not unlike our physicist of the 1930s).

But then he is watching a county match on the last day, dragging on towards an inevitable draw, with a mere handful of spectators. The field is spread out to save boundaries, the batsmen do not bother running for singles. A batsman plays a ball down just in front of himself. To save the fielder the trouble of running in, the batsman steps out of his crease and picks up the ball, intending to throw it back to the bowler. But the bowler has already turned his back on him and started to walk back, so the batsman turns and throws the ball to the wicket-keeper but with his batting gloves on, his aim is poor and he throws the ball into his own stumps. Someone appeals and he is given out. The following day, the score sheet correctly says 'bowled'. Alternatively, as the unfortunate batsman is on the way to the pavilion he is approached by the opposing captain who says something to him and he returns to the crease and continues his innings. (What the captain actually said was, "My dear chap, if we're going to get you out, we want to do it properly, please continue.")

The American returns home convinced he will never fully understand the laws of cricket. One is reminded of some lines of Omar Khayyám:

"Myself when young did eagerly frequent
Doctor and Saint, and heard great Argument
  About it and about: but evermore
Came out by the same Door as in I went."

When it comes to the intangible, invisible topics of electricity and magnetism, analogies are all there is and an analogy is just another name for fairy stories. Yet whilst the latter are never true they may point to a moral and give an insight into a subject that otherwise would be empty and they are therefore most useful. Indeed, they are the engineer's staple diet.

The apparatus shown in Figure 1 consists of a pair of horseshoe magnets mounted on an arm so that they can be rotated close to a copper disc which is free to spin. When the magnets are rotated, the disc appears to be dragged around by an invisible means, but never quite succeeds in attaining the speed of the magnets. The 'explanation' of this force of electromagnetic induction usually takes this form:

"Lines of force from the magnet pass through the disc and when the magnets move the lines cut the disc, inducing an e.m.f. in it. The e.m.f. produces a current which reacts with the magnetic field to produce a force which, by Lenz's law, tries to oppose the cause (i.e. the relative motion) so the disc is forced around in the same direction as the rotation of the magnets."

Detailed examination of this will show it to be nothing but 'mumbo-jumbo', every bit as bad as "Mass is the amount of matter in a body".

Imagine a conversation between a bright pupil of about 17 and a physics teacher offering the above explanation. The latter only gets the first half dozen words out before he is interrupted:

"Sir, what are lines of force?"

"Well, a line of force is the direction in which a free North pole would move."

"Sir, how do you get a free North pole?"

(A short pause follows.)

"Well, actually you can't *have* a free North pole, but if you could, it would be the direction in which it *would* move."

"So they're not real, Sir, these 'lines of force'?"

"What do you mean, 'not real'?"

"There's no chemical formula to them is there, Sir; they're only in your mind?"

(Another pause.)

"Yes I suppose so."

The teacher continues only as far as the cutting lines of force producing an e.m.f.

"So an e.m.f. isn't real, Sir?"

"Oh yes it is. Why shouldn't it be?"

"Because it was produced by something that existed only in your mind, Sir."

"But the e.m.f. produces a current, and *that's* real."

"What *is* a current, Sir?"

"A current is a flow of electrons."

"What is an electron, Sir? Is it really a bundle of waves of probability and therefore not real either?"

(A long pause this time.)

"For the purpose of this explanation you may take the current as real."

(Capitulation without admitting it.)

The teacher goes on to the point where the current reacts with the magnetic field.

"Sir, is this that same set of magnetic lines of force that you admitted only existed in your head?—and *that* produces a *real* force?"

The teacher (head bowed): "Yes."

He is destroyed, and so is the so-called 'explanation'. *It could all be wrong!* But it is all we have at the moment.

Figure 1 A pair of rotating magnets near a copper disc demonstrates induction motor action.

Figure 2 Offset gyro precessing around an 'Eiffel Tower'.

Spin

Electricity and magnetism are one thing, but surely elementary mechanics is beyond question. Here everything is visible and tangible and predictable, and we have Newton's invincible laws to guide us. Yet spin is a topic almost as neglected, scientifically, as is the subject of sleep.

I detect a recent change in school physics teaching whereby 80% of our young people are now taught that there is no such thing as centrifugal force. When confronted by a teacher of this Papal-like edict, I am inclined to remark, "It all depends which end of the string you are holding," and he usually asks for an explanation. I say that when whirling a chestnut on the end of a string in the manner of an old-fashioned aeroplane propeller, I usually hold the end of the string remote from the nut and I feel the nut trying to pull my hand away from the centre. This I define as centrifugal force. You, teacher, are apparently privileged always to be allowed to sit on the nut! So you feel a centripetal force pulling you towards the centre—or don't you believe in action and reaction? If there is a centripetal force on the nut, there must be a reaction on the pivot which is centrifugal. The whole argument is stupid and pedantic.

But when it comes to calculating centri— (whichever you like to call it) force, I have seen teachers use vector product to do it.

They take the angular velocity $\boldsymbol{\omega}$ as a rotary vector normal to the plane of rotation, $\mathbf{v}$ as a linear vector in the plane of rotation, apply the rule of vector products and obtain acceleration as $\mathbf{v} \times \boldsymbol{\omega}$ in the correct direction, of magnitude $v\omega$ which is also correct, as measured by experiment. But they have used the same velocity twice! This method is guilty of knowing the answer and producing a plausible explanation to fit.

One of the golden rules of vector product (never taught, of course) is that you are not allowed to multiply two vectors of the same dimension. There is no meaning to force × force nor to velocity × velocity, nor potential gradient × potential gradient. There is one inglorious exception to this rule (which, of course, is much taught); it is displacement × displacement, which, of course, gives area, of correct size and direction, if you define the direction of an area as being at right angles to its plane, which is the only thing that could have a meaning.

But even displacement is not extendible to a triple vector product, for a rectangular box of sides $\mathbf{a}$, $\mathbf{b}$ and $\mathbf{c}$ has side areas $\mathbf{a} \times \mathbf{b}$, $\mathbf{b} \times \mathbf{c}$ and $\mathbf{c} \times \mathbf{a}$ but the volume is certainly not $\mathbf{a} \times \mathbf{b} \times \mathbf{c}$ for $\mathbf{a} \times \mathbf{b}$ is at right angles to $\mathbf{c}$ and the second product, therefore, always gives a zero result!

The hazards of vector products are even greater than this. You cannot simply take any two vectors and subject them to the treatment of a vector product. What is the meaning of temperature gradient × angular momentum? Both are clearly vectors. Vector product is a very select 'club' where only certain members may be married in this way.

Returning to rotation again for a moment, the objection I raised about calculating centrifugal force by vector product surely cannot be raised about using the technique to calculate the Coriolis force, for a ring free to slide at velocity $\dot{r}$ on a rod rotating at angular velocity $\dot{\vartheta}$ about an axis at right angles to its length involves two quite distinct and separate movements. And the product $\dot{\mathbf{r}} \times \dot{\boldsymbol{\vartheta}}$ gives a vector in the correct direction and of correct dimensions for acceleration. Its magnitude is, of course, $(\dot{r}\dot{\vartheta})$. Alas! The Coriolis acceleration is $2\dot{r}\dot{\vartheta}$. The technique gives the wrong answer numerically.

Vector product is, therefore, simply an *aide-memoire* for those situations where we know it works, as in Fleming's left- and right-hand rules and the rule for the torque/spin/precession relationship of spinning wheels, usually called 'gyroscopes'.

A classical gyroscope usually consists of a wheel mounted in a ring which is itself pivoted about an orthogonal axis to that of the wheel in a second ring which is also pivoted about the third orthogonal axis. These rings are known as the 'inner' and 'outer' rings respectively. An example is shown in Figure 2. It is well known that when a torque is applied to one ring, that ring does not move, but the other ring 'precesses' i.e. moves with constant angular velocity. It is an energyless interchange. The 'input' is a turning force with no motion. The 'output' is motion with no force.

What is fascinating is that the pairs of rings can be operated on quite separately and independently in that torques can be applied to each and the resulting precessions are as predicted as if only one torque were applied at a time. The situation is exactly analogous to the two-axis treatment of electrical machines, where no transformer interaction occurs between coils at right angles (which is why, of course, Fleming's hand rules work for gyros also, although no-one ever told me so).

If now we hang a weight on the inner ring of the gyro shown in Figure 2 at the point of maximum radius, the torque due to gravity will cause the gyro to precess continuously about the vertical axis of the outer ring. If the weight is large such as a kilogram, and the gyro wheel (about 3 inches diameter) is running slowly (about 300 r.p.m.) the precession rate can be as high as 40 to 50 r.p.m. In such an example it is clear that the gyro and its gimbal rings have a considerable moment of inertia about the vertical axis. It is even more obvious that they have an angular velocity about this axis. Thus, it is clear that they must have an angular momentum about the vertical.

But suppose the weight is attached to the inner ring so that it can be instantly detached. As soon as the weight is removed, the precession stops apparently instantaneously! What happened to the angular momentum? The *fact* is that a freely precessing gyro has no angular momentum about the axis of precession for the simple reason that you never gave it any just by attaching a weight to the inner ring.

The same applies to an offset gyro like the toys that are made to precess about the tops of plastic replicas of the Eiffel Tower (Figure 3). This can be demonstrated by making a larger version and allowing the gyro shaft to collide with a movable mass in the course of its precession. The gyro drops like a stone, the struck mass does not move. If this seems a strange result at least it has the support of the higher theoretical realms of mechanics.

But what then of the centrifugal force? Since the most justifiable method of calculating this force is by the rate of change of momentum, and the wheel has no momentum, there should surely be no centrifugal force. Nor is there if the gyro is perfect, i.e. it has all its mass concentrated in a rim of zero thickness connected by spokes of zero mass to a shaft of zero mass. Impossible as this is, very good approximations to perfection can be constructed and the centrifugal force is demonstrably less than 10% of that predicted by classical physics.

It is here that I part company with the latter, for I maintain that the inertial mass of a spinning body changes when it precesses and in a freely-precessing, perfect gyro, it reduces to zero. Moreover, if the gyro is force-precessed faster than its natural rate, its inertial mass becomes negative, as does centrifugal force, so our school physics teachers will at last have to believe in a negative centripetal force, i.e. a centrifugal force of their definition!

A question of dimensions

What justification have I for assuming that inertial mass can change? My reply is simply to ask what evidence have you that in the circumstances I have described it does not?

Let us examine the method of dimensional analysis as applied to the force between electrostatic charges, between magnetic poles and then the gravitational force between masses.

First we have

$$F = \frac{q_1 q_2}{\varepsilon d^2} \tag{1}$$

which written dimensionally in terms of mass M, length L and time T becomes:

$$[MLT^{-2}] = [q^2/\varepsilon L^2]$$

Now we do not know the dimensions of the permittivity ε, so we deduce that the dimensions of q must be given by

$$[q] = [\varepsilon^{1/2} M^{1/2} L^{3/2} T^{-1}] \tag{2}$$

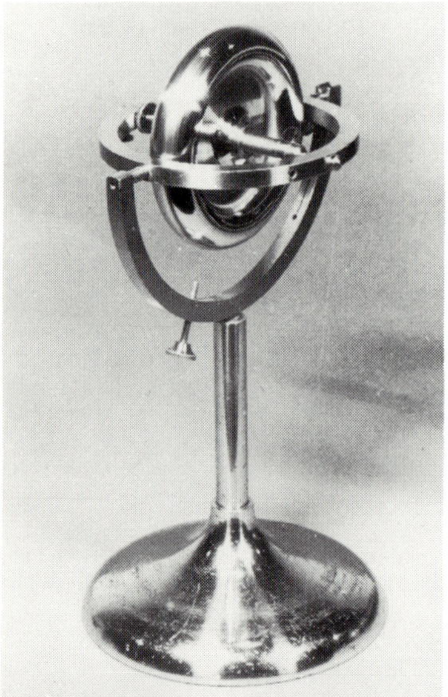

Figure 3 Scientific gyro in gimbal rings.

In electromagnetic studies, the square roots of M and L frequently appear, but never the root of time (strange that?).

If we now repeat the exercise for the force between magnetic poles of strength m_1, m_2, we have

$$F = \frac{m_1 m_2}{\mu d^2} \tag{3}$$

Now equation (3) is identical to equation (1), except that different letters have been used and they represent different quantities, m replacing q and μ, the permeability, replacing ε. Again we do not know the dimensions of μ. It is, therefore, inevitable that our analysis must produce an equation of identical form to equation (2), i.e.

$$[m] = [\mu^{1/2} M^{1/2} L^{3/2} T^{-1}] \tag{4}$$

There would be a total breakdown of logic and the Uniformity of Nature if it were otherwise. Note that in equations (2) and (4) it is the dimensions of the indefinable quantities q and m whose dimensions we sought and were frustrated in that we could not obtain them without recourse to another unknown in each case (ε and μ respectively). I shall call q and m the 'mystic' quantities, simply because you cannot define either except by saying that when you have two of them they repel each other. Not a very satisfactory answer to a child's question "What is charge?" is it? But there always has to be a mystic quantity in science, as there was with the Ancients when earth, air and water were very real but 'fire' was the mystic bit. Do you never get the feeling that someone is not telling us the real truth?

However, let us proceed to the gravitational equation of Newton:

$$F = \frac{G M_1 M_2}{d^2} \tag{5}$$

This is again of identical form to equation (1) with M replacing q and $1/G$ replacing ε (simply because Isaac chose to put his constant on the top of the right-hand side). So we should expect an answer of the form of equation (2). But see what happens in the physics book.

Writing equation (5) in dimensional form, we have

$$[MLT^{-2}] = [GM^2/L^2]$$

and we can cancel an M on each side to obtain:

$$[G] = [M^{-1}L^3T^{-2}]$$

which is a *very* different result from equation (2).

So it must be wrong!

Of course it is. It was wrong to cancel the M because the one on the left-hand side was an *inertial* mass derived from the equation:

Force = Mass × Acceleration

The one on the right was a gravitational mass, the one we should have been regarding as the mystic quantity since it takes the place of q and m in equations (1) and (3) respectively. Let us therefore write gravitational mass as $\boldsymbol{M}$ and try again. The dimensional equation is now very different but identical with the equation following equation (1), thus

$$[MLT^{-2}] = [G\boldsymbol{M}^2/L^2]$$

No cancellation is now possible, we have the same number of unknowns as before (which was not the case with the classical physics approach) and we can proceed as before to find the dimensions of the mystic quantity, thus

$$[\boldsymbol{M}] = [G^{-1/2}M^{1/2}L^{3/2}T^{-1}]$$

and uniformity is restored.

That there is hope even yet, for the next generation of young scientists, is to be seen in a sixth form text book of physics that is becoming popular with the brightest pupils. I refer to the book by Nelkon and Parker[3] who devote separate paragraphs to inertial and gravitational mass, making it clear that when using mass in the force = mass × acceleration sense, it is *inertial* mass that is being considered. They define gravitational mass and give it the symbol m_g (my $\boldsymbol{M}$) and go on to work an example for the period of a simple pendulum, concluding: "Experiments show that to a high degree of accuracy, $T = 2\pi\sqrt{l/g}$ no matter what mass is used, that is, the period depends only on l and g. Thus $m = m_g$, or the gravitational mass is equal to the inertial mass to the best of our present knowledge."

I am bound to conclude with my favourite quotation from Sir Arthur Eddington:[4]

> "However successful the theory of a four-dimensional world may be, it is difficult to ignore a voice inside us which whispers: 'At the back of your mind, you know that a fourth dimension is all nonsense.' I fancy that voice must often have had a busy time in the past history of physics. What nonsense to say that this solid table on which I am writing is a collection of electrons moving with prodigious speed in empty spaces, which relatively to electronic dimensions are as wide as the spaces between the planets in the Solar System! What nonsense to say that the thin air is trying to crush my body with a load of 14 lbs. to the square inch! What nonsense that the star cluster which I see through the telescope, obviously there *now*, is a glimpse into a past age 50,000 years ago! Let us not be beguiled by this voice. It is discredited . . . .

> "We have found a strange footprint on the shores of the unknown. We have devised profound theories, one after another to account for its origin. At last, we have succeeded in reconstructing the creature that made the footprint. And lo! It is our own."

In using the Scientific Method, we dig the bearpits and then fall into them ourselves.

In the last resort, I must return to spin and note that whether we look in the great beyond at the macroscopic, the moons, the planets, the stars and the galaxies, or in the opposite direction, towards the ultimate microscopic—the fundamental particles and antiparticles, all we can ever see is things that spin and if an intelligent school pupil ever asks me now: "What is matter?" I am bound to reply: "My young friend, spin is all there is!"

If they enquire further I can only add: "It is the human mind's concept of rotation (itself a disembodiable thing) which is to be identified with the human mind's concept of matter." You need seek no further. All else is unprofitable.

References

1. Medawar, P.B., *Induction and Intuition in Scientific Thought*, Methuen, London (1969), (1968 Jayne Lecture of the American Philosophical Society).
2. Reynolds, O., *Papers on Mechanical and Physical Subjects*, Volume II, Cambridge University Press (1901).
3. Nelkon, M., and Parker, P., *Advanced Level Physics*, 4th edition, Heinemann Educational Books, New York (1977).
4. Eddington, A., *Space, Time and Gravitation*, Cambridge University Press (1920).

Dr. Graham Cairns-Smith

Graham Cairns-Smith is Reader in Chemistry at Glasgow University. He gained his PhD at Edinburgh for 'Studies in the acridine series' in 1957, having previously achieved First Class Honours in Chemistry there.

His research has covered heterocyclic synthesis, solid state tautomerism, protein binding, oligomer synthesis, enzyme models, clay synthesis, and iron photochemistry. He is the author of two books on the origin of life —'The Life Puzzle' and 'Genetic Takeover'—and of numerous scientific papers.

On the rocks: the vital cocktail?

It is now generally believed that life arose on the Earth spontaneously, that the first systems capable of evolving indefinitely through natural selection were the outcome of normal physicochemical processes. We may accept this as a reasonable premise without being committed to a more particular set of ideas embodied in the doctrine of chemical evolution. According to this doctrine the physicochemical processes in question consisted of a preliminary build-up of our biochemicals (amino acids, sugars and so on) in primordial waters followed by their polymerization and further organization into systems that could eventually reproduce and so become subject to Darwinian selection. This straight line view of the beginnings of organic evolution has been well discussed[1-4] and according to this view organic molecules are the main concern, and indeed those particular organic molecules that now lie at the basis of our biochemistry. Inorganic minerals, insofar as they were involved served in a secondary role, in providing catalysts for the formation of small organic molecules or surfaces on which these might have congregated to make their polymerization more likely to occur. Bernal[5] saw clays in this way and this has been the general view since that time.

It has seemed to me that clays could have had a more direct significance, that they could have been the materials *out of which* the first organisms were largely made. I think that before the means of doing competent organic chemistry had evolved, such central control structures as genes, catalysts, and membranes, could not have been made out of DNA, protein, and lipid. Clay minerals formed continuously in the early environment would have been far more suitable materials.[6-9]

I doubt whether some probiotic soup was needed for the origin of life on the Earth. This idea is widely thought to be predicted by astronomy, consistent with geology, supported by biology, confirmed by chemistry—and perhaps even demanded by logic. But this is not so.

I will consider these points first and then develop the idea that at the beginning of the evolution that was to lead eventually to life as we know it, there were organisms, starter organisms, of a quite different kind.

Background considerations

As a first guess, it was reasonable to suppose that the primordial atmosphere of the Earth would have been strongly reducing, like that of Jupiter, if the Earth formed from a cold cloud of gas and dust that was predominantly hydrogen. The original experiment of Miller[10] demonstrating the easy formation of four of our most fundamental amino acids was designed on this basis. Most current theories of planetary evolution, while agreeing that the Solar System had a cold start, take a more complex view of subsequent events. The first atmosphere was not simply part of the original gas cloud, nor was it necessarily strongly reducing. Walker,[11,12] for example, sees the early atmosphere as faintly reducing at most by the time conditions had settled down sufficiently for life to be possible and proposes a primordial atmosphere like our atmosphere now with the removal of all oxygen and the addition of at most a few per cent of hydrogen.

Indeed, the recent trend has been to see the early Earth as hotter and more dynamic than previously—and hence, running more rapidly towards its present condition. Moorbath[13] suggests that a modified doctrine of uniformitarianism is still tenable as far back as 3.8 billion years—the age of the oldest known metasediments. These contains small amounts of graphite and plenty of carbonate,[14] suggestive of CO_2 rather than CH_4 in the atmosphere.

Clearly, in such an uncertain field, we should not be dogmatic: but in line with the scepticism of Abelson[15] and Sillen[16] the popular idea that life arose under a methane-laden atmosphere with the seas rich in organic molecules has little support from either astronomy or geology. The atmosphere may well have been neutral—mainly N_2 and CO_2. In that case, organic molecules might still have formed locally, from CO_2 under the action of high-energy ultraviolet light penetrating an anoxic atmosphere. But that same radiation would mitigate against the build-up of large 'food supplies': it breaks as well as makes organic molecules.[17]

At first glance molecular biology might seem to support a straight line hypothesis. All organisms now alive have a similar, quite small set of small organic molecules as fundamental units and some common source is strongly suggested. It is clear from the detailed similarity of all present organisms (in the genetic code, the amino acid set etc.) that the source was a common ancestor. The sophistication of this ancestor implies that it was quite high up the evolutionary tree (see Figure 1*a*). We can say also that it had already settled on a *modus operandi* which subsequent evolutionary processes were unable to change.

This is not surprising. Features in organisms tend to become fixed when they have been present for a long time—presumably because later changes presuppose them. But that is not to say that central biochemistry was always fixed; any more than that the present invariability of hair in mammals says that hair was always in our ancestry. (All we can say here is that hair was probably already frozen-in in the common ancestor of the mammals—a beast that had, nevertheless, non-hairy antecedents.) Similarly it is too hasty to conclude that first life had all or any of the biochemicals that all life now shares (see Figures 1*b* and 1*c*).

There is no doubt of the importance of Miller's experiment and of the subsequent work on the abiotic synthesis of biochemicals that it inspired.[18,19] The variety of conditions found to be fruitful, together with the discovery of biochemicals or their precursors in such remote places as meteorites[20] has made it plain that our biochemistry is based on a set of molecules that are easier to make than one might have supposed. But what is the significance of this? That our biochemicals were formed in the early environment and then used as building blocks for first life? Not necessarily. One might as well argue that because the bone material apatite is a mineral that would have been in the environment before there were bones, then the first animals with bones must have picked them up from the environment. Convergences of this kind may simply reflect the generality of the laws of chemistry. What happens in organisms may be similar to what happens elsewhere, but not because one occurrence causes the other. An evolved biochemistry might be expected to be based on a limited number of readily made stable components. That would be good engineering, an understandable outcome of natural selection. But it is not at all clear exactly when the selection of our particular set of biochemicals was made.

In any case, the success of the simulation experiments has been, at times, overstated. The products are invariably complex mixtures, usually largely polymeric tars. Furthermore some critical molecules of intermediate complexity, particularly nucleotides and lipids, have not been shown to be formed under simple uncontrolled conditions.[3,18]

There is a simple-minded logic that says that to make a machine you must start with the components. But evolution is different: all that we can say here is that evolution had to start with *some* machine that was capable of evolving indefinitely through natural selection and that the components of *that* machine had to be supplied by the environment. But did the machine that could most easily start have to be similar to the machine that would be preferred eventually? It is a question of how radical very early evolution could have been.

Takeovers in evolution

Was evolution at the biochemical level always comparable to the changes that took place when (say) the fins of fish transformed into the limbs of land animals: was there a continuous line of transformations, that is, with functions sometimes changing but structures remaining related? Or did

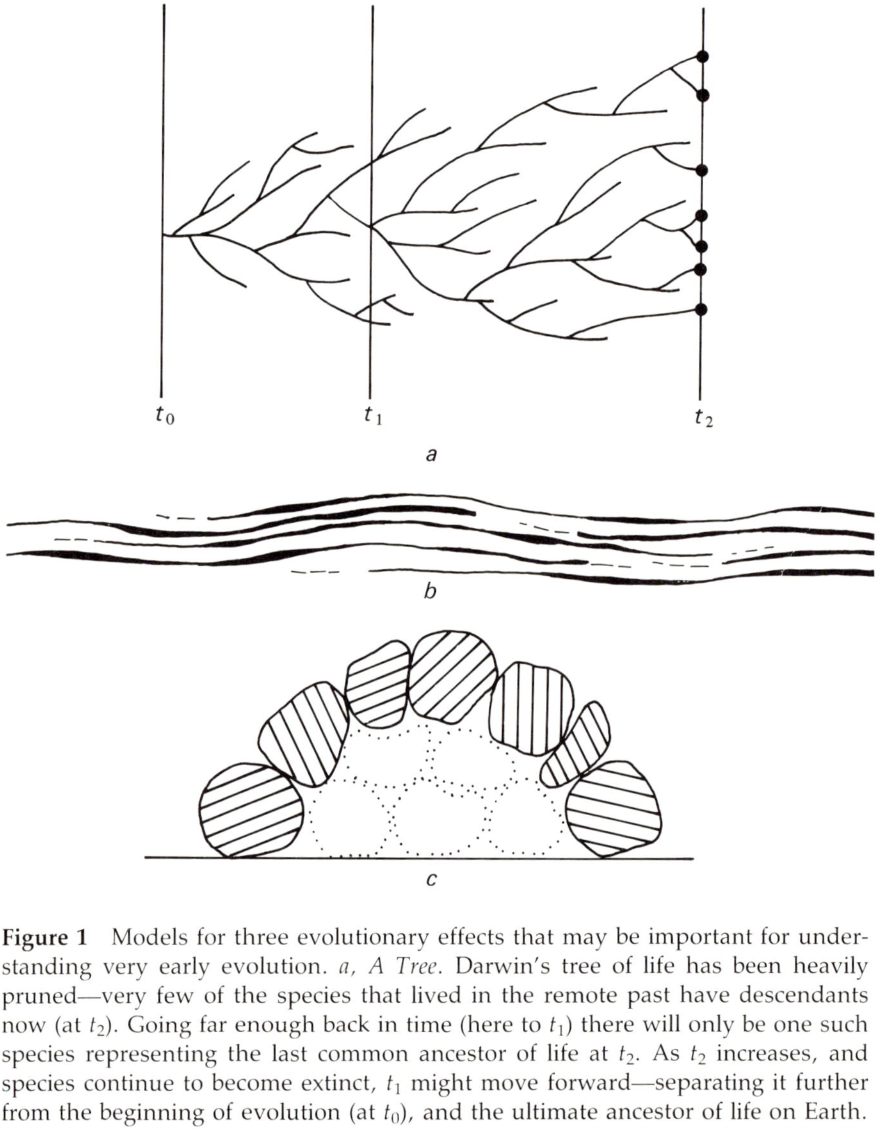

Figure 1 Models for three evolutionary effects that may be important for understanding very early evolution. *a, A Tree.* Darwin's tree of life has been heavily pruned—very few of the species that lived in the remote past have descendants now (at t_2). Going far enough back in time (here to t_1) there will only be one such species representing the last common ancestor of life at t_2. As t_2 increases, and species continue to become extinct, t_1 might move forward—separating it further from the beginning of evolution (at t_0), and the ultimate ancestor of life on Earth. That the distance between t_0 and t_1 is now very considerable is indicated by the great sophistication of the machinery that is common to all life. *b, A Rope.* The lines in (*a*) are multi-stranded, because an organism consists of a collaborating set of subsystems. Thus while an extended line of organisms may be continuous—that is, with its members related through a sequence of minor modifications—the subsystems need not be: old subsystems (e.g. gills) may 'peter out' while new ones (e.g. lungs) 'peter in'. The strands may be shorter than the rope, that is, with those at one end discontinuous with those at the other—they might indeed be made of

early evolution sometimes proceed through takeovers, with functions sometimes persisting while structures changed discontinuously? This would be comparable to the change from gills in fish to lungs in land animals. Here a vital function was maintained through transformations that disconnected earlier and later means: first there were gills; then there were bimodal animals, with both gills and lungs (the latter derived from the alimentary canal); then, in many animals, there were only lungs. The whole process was gradual, but that is not to say that there was a succession of structures intermediate in form between gills and lungs. Evolution through takeovers requires no structural similarity between the first way of performing functions and later ways. Continuity between organisms in an evolving line does not imply a long term continuity of the subsystems that make them up. We might compare a line of organisms with a rope which may be continuous for a hundred metres although its individual fibres are only a few metres long (Figure 1*b*).

Clearly logic does not exclude the possibility of takeovers in evolution, since they happen. We can think of an extended line of organisms as a kind of evolving technology in which new materials and manufacturing methods are discovered. The discoveries are made gradually and in the first place in response to selection pressures operating on existing functions.[21] Bone presumably evolved in the first place as a structural material. Once organisms had acquired the necessary manufacturing methods, bone was found to have other uses—for example, for sound transmission in the mammalian ear. Keratin is a similar story: appearing in early land animals as a material for water retaining scales, it came to be used for claws, feathers, hair etc.[22] Such discoveries sometimes mean that a new material or a new technique makes an old one redundant.

---

different materials. Similarly, during early biochemical evolution a coming and going of subsysems could have disconnected the original from the final materials out of which organisms were to be made. *c, An Arch.* The invariance of our central biochemistry—established by t_1 in *a*—is largely due to its subsystems leaning on each other: everything depends on everything, more or less. The whole system is much too complicated just to have happened; its present rigidity must be a consequence of an extended evolution. The problem here is similar to that of how an arch of stones might be built through a series of minor modifications—adding, subtracting or adjusting one stone at a time. This problem can be solved with some kind of scaffolding of stones which are subsequently removed. The strong interdependence of the subsystems in our biochemistry is *prima facie* evidence for earlier subsystems now lost. There is no particular reason to suppose that the missing subsystems were made the same way as any of the remaining subsystems, and indeed reasons are given in the text for supposing otherwise.

Now consider the problem at a deeper level. Our last common ancestor already possessed the means of manufacturing protein which is the central multifunctional material of modern organisms. But before organisms had invented the means of putting together amino acids into particular genetically-controlled sequences, would amino acids have been so generally useful? And what functions of poly-amino acids provided the selection pressure for the evolution of the equipment for protein synthesis? It is not to be assumed that this function was catalytic because enzymes now are so important. That would be like supposing that keratin was invented so that birds could fly. That is not the way evolution works: it does not see into the future. From the intimate dependence of an enzyme's function on the way its amino acid chains fold, and from the dependence of that on accurate sequence specification, it is more likely that some much less taxing function came first. Once accurate manufacturing machinery was there, this new stuff protein could demonstrate its versatility and in doing so, displaced other more pedestrian materials. What the immediate precursors of protein were is speculative, but that there were such precursors is I think much less so. Before there was protein there were surely other means to its now vital functions, as before there was steel there was bronze and before that a stone age (see ref. 23).

Nor does logic demand that a genetic material be immune from takeovers. A first genetic material, let us say, can form spontaneously in non-biological surroundings. This is the basis of primary organisms which evolve a technology within which more sophisticated genetic materials become possible. One of these forms a second class of genes in the now genetically bimodal organisms. The secondary genes turn out to be more efficient and eventually come to control everything required for their own production. The first class disappears (see Figure 2).

The solution to the problem of the origin of life can only be impeded by insisting, in the name of Ockam, that only the pieces now visible on the chessboard are to be used. This ignores a well known and easily understood feature of evolution—it subtracts as well as adds. It is unlikely that the components most suitable for a first system would still be retained by an advanced system—given that a succession of additions and subtractions provides a means of radical change. Evolution had no prior knowledge of which molecules to choose initially to make the best machines in the future: in particular it was not to know that organic molecules—so difficult to control—would be ideal once means of control had evolved. Life would start, not with a predestined set of official biochemicals, but with whatever available units could most easily put themselves together into the kinds of hardware needed for a system with evolutionary potential—genes, catalysts, membranes and such things. We seek the missing pieces of a stone age biochemistry (compare with Figure 1*c*).

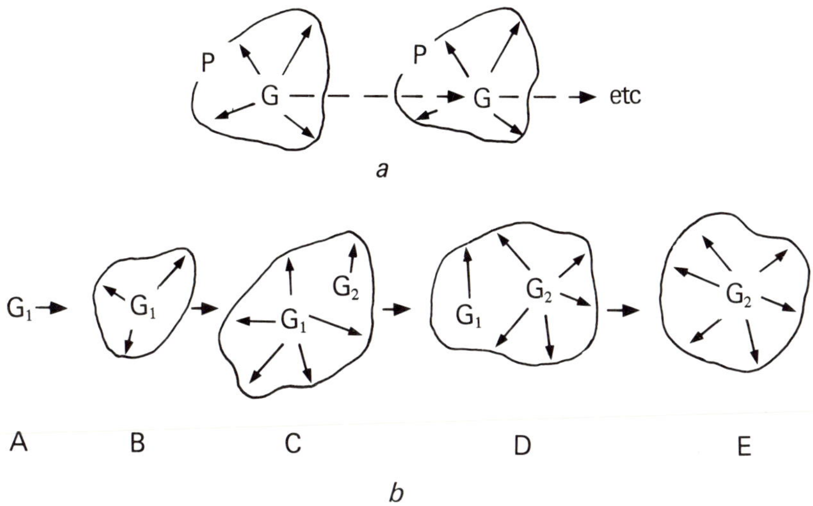

Figure 2 *a*, An organism consists of a genotype and a phenotype. The genotype is a set of instructions—like a musical score. It is a set of particular patterns (genes) written on a special kind of paper—a genetic material. The genotype has two properties: (i) It can be printed by a templating process (dashed arrows) and (ii) It specifies the production of a surrounding phenotype (solid arrows). The phenotype is the expression of the genetic information and its function is to provide an environment that protects, and assists in printing and publishing the genes. The problem of the origin of life is largely that a very elaborate phenotype seems necessary to operate the present genetic material, DNA. (This score can hold a lot of information, but it needs sophisticated equipment to print it or to read it). *b*, A general solution to the above problem—genetic takeover. Here we suppose that a first kind of genetic material G_1 could operate without any phenotype (stage A). These primary organisms spread to fill increasingly difficult niches by evolving the means to specify a surrounding phenotype (stage B). A secondary genetic material, G_2 appeared later within such a phenotype on which it was wholly dependent (stage C). In subsequent genetically bimodal organisms, the second genetic material proved to be more efficient and versatile, and eventually displaced the starter material (stages D and E).

A clay-based primary life?

Millot[24] has reviewed the salient features of clay structure and genesis. A typical clay crystallite consists of two-dimensional polymers—unit layers—stacked on top of each other. The individual unit layers are made up of either three or four sheets of oxygen atoms (formally as O^{2-} and OH^- anions) with small cations, especially Si^{4+} and Al^{3+}, nested in (tetrahedral and octahedral) sites between these oxygens. In the very common kaolinite structure, for example, there are three sheets of oxygens in the unit layer with one sheet of silicons and one of aluminums between them. Here the arrangement of O^{2-}, OH^-, Si^{4+}, and Al^{3+} is such that the charges cancel exactly. A typical kaolinite crystal is a neat stack of neutral unit layers. In other clays, such as montmorillonite, there are three sheets of cations between four sheets of oxygens and hydroxyls. Here the charges do not balance so that the unit layers have an excess of negative charge that is compensated by larger cations, such as Na^+ and Ca^{2+}, between them. There is also commonly water between the layers of such clays. Unlike the cations within the layers, the inter-layer cations are quite mobile and can be exchanged by others—including organic cations. Water too can be partly or wholly replaced by organic molecules, and indeed most kinds of clays allow organic molecules of one sort or another between their layers.[25]

Generally, montmorillonite layers are much less neatly stacked than the kaolinite type. Montmorillonite often has the appearance of crumpled linen under the electron microscope. The individual layers are flexible membranes. Individual kaolinite layers are flexible too, being swiss-rolled into tubes in the mineral halloysite.

The more recently discovered clay mineral imogolite consists of seamless tubes that have an internal pore of about 1 nm (ref. 26). The related allophane consists of tiny (3.5–5.5 nm) porous hollow spheres.[27]

So if it were such things as membranes, tubes and vesicles that were needed for the first organisms, the clay minerals could most easily have provided them, the clays that we have been discussing form continuously in the environment from solutions derived from rock weathering.[24] Presumably this would have also been so on the primitive Earth.

Of course, a heap of tubes and membranes is not an organism. For that there must be organization under genetic control. There must be a long term genetic memory, presumably through a genetic material of some sort that could hold and replicate specific more or less complex patterns (Figure 2*a*). Are there clays that could do that? If so we might have a lead to the true beginners at evolution (G_1 in Figure 2*b*) which, unlike DNA, could work without preexisting machinery.

It is not difficult to see how clay crystals might hold large amounts of information.[8,9] For example, there is often a considerable freedom in the ways in which layers may be stacked on top of each other. Hence, rather as DNA holds information like a (one-dimensional) message tape, as a particular base sequence; so a clay crystal could in principle hold information as a particular sequence of stacking modes. This would be a one-dimensional crystal gene. Here we will consider another kind of crystal gene, a kind that would hold information as in a punched card, as a specific two-dimensional array of features.

Two questions arise: (1) How could two-dimensional patterns be reproduced; and (2) how could the patterns mean anything?

The replication requirement

If during crystal growth a particular arbitrary feature in the arrangement of cations in a preexisting layer is copied in the next layer growing on top of it, then the resulting crystal should appear disordered to some extent *within* the layers—a particular form of the disorder representing the 'information'—while at the same time ordered *between* the layers. One approach to the search for a clay genetic material, then, is through the search for inter-layer ordering in clay crystal structures. Although a difficult field of study for X-ray crystallography, there is evidence for such inter-layer ordering in several kinds of clays.[28]

Another approach to the search for clay genes is through synthesis. Weiss[29] found laboratory evidence for the replication of features of cation substitutions during clay synthesis. He found that a number of properties that depend on silicon atoms being replaced by aluminium atoms in synthetic clays can be inherited by successive seeding over several generations.

Here I suggest a third approach—through the study of crystal morphology. Figure 3 illustrates what crystal genes might look like that held information in the form of a specific two-dimensional pattern of some sort that was replicated during growth in the stacking direction. A columnar or tabular morphology is to be expected since the sideways extent of new layers must not either grow or diminish as the crystal grows. This would either add 'nonsense' information or lose information, respectively.

In its outer form at least, kaolinite often has the general sort of columnar appearance idealized in Figure 3*c*, when it occurs as 'booklets', 'concertinas' or 'vermiforms'. These crystals may resemble a tall stack of paper, in which each sheet has the same or a similar rather complex shape. As a result the columns are grooved or fluted. Many of these are shown in the electron micrographs by Keller (ref. 30 and references therein). Long vermiforms, such as those in Figure 34 in Keller[30] may give the appearance of having been extruded. But it seems more likely that the constant cross-section arises from growth having been exclusively in the stacking direction, as required for our putative genetic crystals. If this is so then

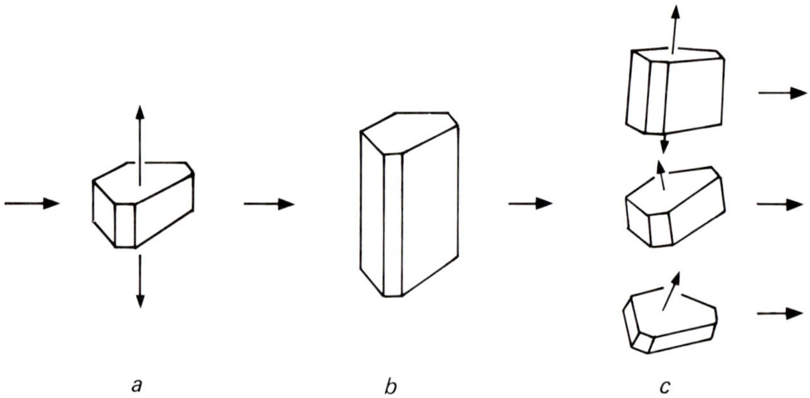

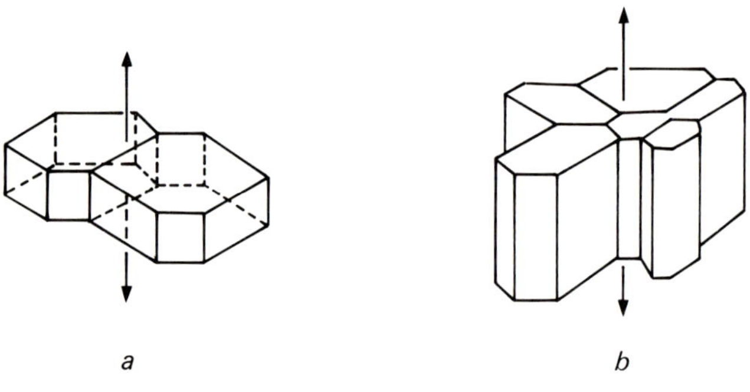

Figure 3 A crystal gene that replicates information written in two-dimensions might be expected to have a tabular or columnar morphology. *a* shows a replication cycle consisting of (i) growth (exclusively) on two faces that are in the plane in which the information is held, and (ii) cleavage (exclusively) in this plane. The information might simply be the constant shape of the growing faces. This could be increased if the crystal was twinned and if the planes of intersection between its constituent domains remain parallel to the sides of the crystal (*b* and *c*). The information might then consist of a particular crazy-paving pattern of domains rather as the information in a punched card consists of a particular arrangement of holes. Crystal genes with information in one dimension can also be imagined.[8,9]

'information', in the form at least of the constant shape and size of the layers, was replicated quite accurately as the crystals grew.

Fluting and grooving of crystals is an indication that the crystals are twinned, made up of domains as in Figures 3*b* and *c*. In that case, more complex information might be reproduced during crystal growth, not just the shape of the individual layers but a particular crazy-paving patterning of domains written in each of the layers. (For our discussion it is not necessary to know the underlying cause of the twinning, but it seems likely that it arises from there being domains with alternative arrangements of the aluminum cations within the layers—as in the pseudo-twinning discovered by Mansfield and Bailey[31] in much larger kaolinite crystals.)

At present, all we can say is that it is crystallographically plausible to suppose that during a layer-on-layer growth process specific defects in the layer below may be copied into the newly forming layer—and that there is some evidence for such effects. So let us move on to the next question. Even given such a spontaneous printing mechanism, could what was being printed have any significance—what could the information mean?

The control requirement

Genetic information tends to preserve, replicate and propagate itself. Those messages are selected that can control their immediate environment to this end. For all life that we know of on Earth, the means to this end are immensely complex. Many intricate DNA messages must collaborate before any of them make sense—before they give rise together to an efficient vehicle, an organism, that ensures their propagation into the indefinite future. The messages in the first genetic material(s) must have had a much more direct meaning. Considering the kaolinite model, how could a message in the form of the particular shape and size of the replicating cross-section of a mass of columnar crystals tend to be preferred?

The answer to this question is: Quite easily. Many large-scale properties of materials are controlled by the shapes and sizes of their constituent particles. Soil chemists are familiar with the idea that rheological properties and the general porosity of soils are controlled by the shapes and sizes of clay and other particles. We might imagine clays forming in a sandstone from silica, alumina etc. dissolved in percolating waters. The mechanically protected pores in sandstone provide excellent reaction vessels—and indeed most sandstones contain clays (kaolinite and others) that have grown *in situ*.[8] A clay will only continue to grow if it does not clog up the pores. A replicating clay, then, would be more effective, and the message it was replicating would become more prevalent, if that message consisted of a particle shape and size that gave rise to the appropriate porosity. Fraser[32] has a more subtle idea, that crystal shapes and sizes would have been selected that were (literally) in tune with vibrations in the environment. We would be justified in calling such a message genetic in the biological sense because it would be making things right for its own propagation. (That is what DNA messages do if now in a very indirect way.)

Where there were different kinds of replicating crystallites each would soon create a local zone containing millions and millions of copies. Because of effects of the replicating defect patterns peculiar to each zone, some zones would be more suitable for clay synthesis than others. Some zones would grow faster, like competing colonies of bacteria in a Petri dish where the faster reproducers have made bigger patches by the time nutrient or space runs out. But evolution depends on such competitions continuing indefinitely. It depends on there being environments that are open systems (chemostats not Petri dishes). Clays are very often formed in open systems—in sandstones for example—where endless supplies of nutrient weathering solutions percolate in, and where continuously or intermittently, clays are being carried away. Under such circumstances, according to the simple mathematics of the chemostat,[33] a marginal difference in growth rate can lead to a faster replicating mutant displacing the other *entirely*.

On the medium term, then, one might imagine replicating clays evolving like this within a given initial environment that was generally favorable to clay synthesis. Through successive replacements, more efficiently replicating types would appear, i.e. clay defect patterns would appear that contrived their surroundings better to favor the replication of clay defect patterns. On the longer term we could predict another trend. Clays that had evolved in a first easy environment would spread to adjacent, more difficult environments where clays could not previously have formed. In such places only evolved clays would be able to propagate. By that time you could hardly deny them the title of organism.

Sometimes kaolinite indeed seems to make things right for its own synthesis: booklet kaolinites have a particularly porous texture.[34] Not only should a porous texture favor clay synthesis by allowing an adequate through-put of nutrient solutions, but a high porosity seems to be favorable to kaolinite synthesis particularly.[35] Such positive feedback effects were suggested by Clemency[36] in a case where adjacent kaolinite and montmorillonite each appeared to be maintaining conditions suitable to their own synthesis, by controlling the concentrations of ions in their surroundings. Such an effect might conceivably be a consequence of a natural selection of a very simple sort: from an initial population of crystallites those that tended to give rise to booklets were self-selecting because they continued to maintain suitable conditions for crystal growth while at the same time propagating their kind. If this were so, then sets of booklets in a given micro-region that had been derived from common seeds should show detailed morphological similarities, of groove arrangements and cross-sectional area, for example. Furthermore, if they have been the products of a multistage mutation and selection process they might appear to be contrived in a more detailed way, for example, the exact size and shape might be particularly suited to maintaining the conditions for kaolinite synthesis in a particular locality. Hence perhaps the very beginnings of evolutionary processes might be discoverable on Earth today.

I think, then, that kaolinite might be able to evolve under natural selection: superficially at least it seems to have an appropriate crystal structure, it can have an appropriate morphology and it can form under suitable conditions. Perhaps, it sometimes shows signs of having evolved. It is too soon to attempt to identify the primary genetic material(s), but we can say perhaps that kaolinite is a model for several of the features to look out for in clays.

The test question for any systems that may have been the products of natural selection is this: Do they reproduce and does their structure appear to be contrived so as to increase the efficiency of that reproduction? 'Contrivance' is hard to define but may be easy enough to recognize if one is on the lookout for it. One can recognize human artifacts, like axe heads, because they belong to a small class of forms that turn up more often than you would expect on the basis of chance and normal physical processes—and because they can be seen to have a function.

Evolutionary potential of clays

We have seen that a genetic material might itself evolve insofar as information that it contains may affect its physical properties, and hence, in suitable circumstances the propagation of the information. Clearly this direct approach is limited: there is a limit to the things that (say) kaolinite could do directly, however intricate its defect structure might be. One might conclude, indeed, that the evolution of a clay could only be trivial. But that would be too hasty. In principle, indirect modes of control could give significance to specific intricate genetic patterns. One may see it indeed as a characteristic of the evolution of genetic control that it tends to become increasingly indirect. Given the necessary reading equipment, the most abstract-looking messages can have a meaning. No, the intrinsic properties of a genetic material, the properties that it acquires immediately by virtue of information that it contains, although important to give it a toe-hold, are not the main factors in deciding whether, in principle, that genetic material could sustain a long-term evolutionary process. For this there are only two absolute requirements: the genetic material must be able to hold and replicate accurately large amounts of information; and it must come to control its environment indirectly.

Suppose that a clay unit layer with N unit cells consists of n domains. For reasonable values of N and n there can be an enormous number of possible crazy-paving domain patterns. Suppose for simplicity that the position of each domain is determined by one critical feature (for example, the position of a central screw dislocation). For $N >> n$ there would be $\sim N$ possible positions for each of the n features, that is N^n in all. A crystallite of $1 \mu m$ across containing 10^6 unit cells per layer and 100 domains would thus be one of about 10^{600}. The crystallite could hold about as much information, that is, as a large protein molecule. To be retained from generation to generation the crystal growth processes would have to be very accurate. Assuming that, the next question is how the information might be expressed indirectly.

Indirect techniques of control

DNA replicates genetic information and protein does something about it. Although not essential, it is clearly a good idea to have specialist structures for what are very different activities—of replication and control. Being more complex and not being absolutely essential, such an arrangement is not to be expected initially: but being more efficient it is perhaps a predictable outcome of evolution.

For a crystal to be a good replicating machine it should probably be rather rigid with strong forces between the units so that few mistakes are made. Some typically well crystallized material like kaolinite, vermiculite or chlorite was perhaps the original counterpart of DNA. To be a versatile control structure, however, it is better to be flexible, like protein, so that information within it can make a variety of complex forms. To evolve very far, then, we might guess that an original clay genetic material would come to operate through more flexible intermediate structures analogous to protein. Organic polymers, perhaps even peptides, are a possibility given environmental monomer supplies. (Lahav *et al.*[37] have shown that short chains of amino acids can be made on clays). But I think that peptides would only have become a good idea much later: clays are not especially good at binding amino acids, nor at joining them together. To begin with there would have been easier ways of making secondary flexible control structures—by controlling the formation of other clays like halloysite or montmorillonite.

Suppose that under conditions A, crystal growth of a genetic material is favored, but that under conditions B—a different temperature or pH perhaps—another clay is favored that grows epitaxially on the genetic crystals. The way the second clay grows depends on the surface characteristics of the first and these in turn depend on the domain structure or other form of genetic information in the first clay. The second clay is then a kind of phenotype for the first—if the success of the first depends to some extent on the characteristics of the second. Perhaps, because of the secondary clays surrounding them, the first clays will grow faster when conditions A return, or will be protected against periodic inclement conditions C.

We are imagining now a situation where instead of a genetic clay making conditions right for itself directly, it makes conditions right for other clays, which in turn make conditions right for the genetic clay. This roundabout approach would have the advantage that the secondary clays would not have to be able to replicate the information that was partly responsible for their formation. Freed from this constraint there would be a wider range of possibilities open. The secondary clays need not be hard crystals; they could be flexible gel-forming membranes, or tubes or vesicles.

Organic chemistry without enzymes

In the absence of the detailed microscopic control that the active centers of enzymes can provide, there has to be more control on the larger scale. Solutions must be mixed, conditions adjusted, products separated, etc. Each operation may be simple in itself, but for a multi-step synthesis the level of contrivance required to create an appropriate sequence of operations is very considerable. Quite simply, specific organic molecules are usually difficult to synthesize. Yet before there were enzymes there must have been some means of doing organic chemistry: this must have become very effective to create consistent clean supplies of nucleotides and other components required before our protein synthesizing machinery could have even begun to evolve. I think that before there were enzymes, there was an altogether different kind of biochemistry that depended much more than our life does on genetically specified apparatus. Our life is protein life: biomolecular control depends absolutely on the unique socket-forming abilities of proteins, on their ability to recognize other molecules. Thus many reactions can take place in the same vessel without confusion. Once this ability had evolved, biochemistry would never have been the same again. But we must try to look back beyond that technological revolution. When you do not have the means of recognizing individual molecules, how do you keep control? I can see no alternative to having physically laid out production lines of some sort, with tubes, pumps, mixing chambers and so on. Then molecule A reacts with molecule X, not through some structure that recognizes both, but through some kind of 'glassware' that localises both. Of course, there is nothing probiotic about this: apparatus of any complexity could only be the result of contrivance, the product of evolution through natural selection.

In ordinary clays we have some likely-looking raw apparatus. (There is even a proton pump that works on a thermal gradient across finely divided $Al(OH)_3$, $Mg(OH)_2$ or $Ca(OH)_2$.[38]) And there are intricate interlayer compartments, catalytic effects at special sites—and so on. Could natural selection operating on the kinds of clay-based organisms that we have been imagining have put suitable apparatus together? Can we imagine at least the very beginnings of such construction work?

CO_2 might have been a sufficient carbon source for the first essays in organic chemistry. CO_2 can be reduced in aqueous solution to small organic molecules (in particular formate, formaldehyde and oxalate) by the action of UV light.[39–41]

Now imagine an assemblage of genetic and secondary mineral crystallites exposed to UV sunlight. Adsorbed CO_2 is partly converted to a mix of organic molecules. These further react. Often they are photolysed back to CO_2. The steady-state concentrations that are arrived at will depend on complex factors including physical characteristics of the mass of crystallites—veriforms, membranes, tubes etc. This mass creates a richly heterogeneous set of microenvironments; some protected from the UV light, others tending to bind certain types of organic molecules, still others containing catalytic sites.

Ions of transition metals, like iron, might have been particularly important, both free and incorporated into clay layers. Such ions would be UV-shields but they would also themselves have photochemical effects. Radiation in the 200–400 nm range can readily generate oxidising and reducing agents—Fe^{3+} from Fe^{2+}, for example, and radicals such as H.[41] Where transition metal ions are incorporated asymmetrically in a clay membrane one might expect asymmetric effects—protons, hydrogen atoms or solvated electrons being released preferentially on one side of a clay membrane. Thus chemical potentials could be generated, further complicating the situation. There would be microenvironments of different *p*H, *p*E etc., and inevitably flows generated between them. All these, at first, quite fortuitous effects would influence the yields and locations of different organic species. Let us concentrate attention on one—oxalate.

Oxalate is one of the species recognized by Getoff as a product of inorganic UV photosynthesis. It is also easily destroyed by UV in the presence of Fe^{3+}. Oxalate concentrations might thus be expected to be particularly sensitive to the kinds of factors that we have just discussed. Oxalate binds aluminium and is also a catalyst for kaolinite synthesis.[42]

Let us suppose, then, that kaolinite could have been a genetic material. We suppose that domain patterns in a population of replicating kaolinite crystals control the morphology of the crystals themselves and, through epitaxy, the shapes, sizes and charge distributions of secondary clays. The genetic information thus exerts a remote and partial control on the curling, folding and interparticle interactions of the secondary clays and hence on the architecture of the maze of pores and interconnected compartments that any clay mass inevitably makes. If the net effect of all this is to increase slightly the oxalate concentration surrounding one particular set of replicating crystals with one particular domain pattern, then that pattern could become more prevalent with time. In line with the earlier discussion on the advantages of a suitable porosity, if a genetic pattern in any way contrives its immediate environment to make or accumulate organic molecules that are favorable to clay synthesis then that pattern is fitter in the biological sense: it has more offsprings more quickly. A chance assemblage of pores, compartments and so on might bring about some marginal effect. In that case, apparatus could already be under construction by natural selection with further improvements to be expected. The limit to such improvements would be set by the amount of genetic information being accurately printed. That might have been very considerable.

We might predict that metal binding agents generally would have been important components of early phenotypes that were creating conditions to encourage the replication of an inorganic crystal gene—whether kaolinite or not. Krebs acids and amino acids are metal binding agents. Perhaps these were present in primary phenotypes for that reason and hence became incorporated into the system that was to take over from it.

However the competence was evolved to make amino acids and then nucleotides, the primary organic biochemistry looks as if it started from CO_2 and first made mainly C–H–O compounds—because these are still at the heart of our biochemical pathways: nitrogen almost looks like an afterthought.[7,43,44] With a clay-controlled organic biochemistry, nitrogen *could* have come in late. In any case, the reactivity of amino groups would have made organized organic chemistry difficult until suitable means of handling them had evolved. (As Abelson[15] points out, amino acids and sugars, for example, tend to react with each other to give complex brown products.) The simplest way of imagining the eventual introduction of nitrogen would be through the evolutionary discovery of N_2 fixation. Schrauzer and Guth[45] found that $Fe(OH)_2$ can convert N_2 to ammonia and hydrazine, the reaction being promoted by UV light. Nitrogen fixation should thus not have been beyond the devices of wholly inorganic control machinery.

There need never have been external soup, then, to provide 'nutrients'—nitrogenous or otherwise—for first life. I suggest that different kinds of organic molecules were made locally in the phenotypes of early clay-based life—when they could confer a selective advantage and when their synthesis became possible. Perhaps, the ground plan of our biochemical pathways was gradually sketched with the aid of, eventually, some very exotic clay machinery. (Perhaps some signs of this machinery might be found in the most ancient sediments.) Only much later, within the phenotypes of some line of evolved primary organisms, were the first nucleic acid-like molecules invented to set the stage for the beginnings of a new kind of genetic control machinery (Figure 2). But that is another story: the true beginnings of organic evolution were, I think, in a different part of chemistry.

Conclusion

From its remote origins organic evolution has proceeded through continuous successions of organisms; but that is not to say that the subsystems in organisms now are related continuously to initial subsystems. Through takeovers, functions may be maintained while the structures carrying out these functions change (the rope analogy, Figure 1b). Thus there is no overriding restriction imposed by current biochemistry on the material nature of the first means of carrying out functions such as the transmission of hereditary information.

Indeed, we would not expect that the materials most suitable for first life would have continued to be so as techniques of fabrication evolved. To begin with, selection would favor materials that happened to organize themselves into appropriate forms: later, materials would be favored that could *be* organized. In essence this is why I think organic evolution (organic in the biological sense) would have started with inorganic materials (inorganic in the chemist's sense). Many inorganic polymerizations—for example, the formation of clays from dilute solutions—are simultaneously crystallizations. (this is not so for organic polymerizations.) Local thermodynamic control can thus correct errors: when crystals are growing slowly enough a wrong ion adding will usually, sooner or later, come away again. The process is slow, but it can lead to a highly organized polymer. Fortuitously such polymers are often membranes, tubes etc., and they might have had an early biochemical use. Even that most difficult and critical of functions—the printing of genetic information—may well be a fortuitous effect in certain growing clay crystallites. So perhaps that is the way that life could have started. But it is not the way it would have gone on, given that organic molecules became involved.

Organic materials are potentially more able to *be* organized. Once put together correctly, two carbon atoms may stay together indefinitely, so there is much less scope for automatic error corrections: mistakes stay put too. Generally it takes a skilled organic chemist, or an enzyme, to avoid a tarry chaos. Thus, the statement that organic molecules are ideal for building biomolecular machinery is only partly true. Organic molecules are ideal for life in the same way that a violin is ideal for making music—in the hands of an expert who can control the richness of possibilities. Organic molecules are ideal for evolved life, but first life was not to know this.

References

1. Oparin, A.I., *The Origin of Life on the Earth*, Oliver and Boyd, Edinburgh (1957).
2. Calvin, M., *Chemical Evolution: Molecular Evolution Towards the Origin of Living Systems on the Earth and Elsewhere*, Clarendon, Oxford (1969).
3. Miller, S.L., and Orgel, L.W., *The Origins of Life on the Earth*, Prentice-Hall, New Jersey (1974).
4. Day, W., *Genesis on Planet Earth*, 2nd Edn, Yale University Press, New Haven (1984).
5. Bernal, J.D., *The Physical Basis of Life*, Routledge and Kegan Paul, London (1951).
6. Cairns-Smith, A.G., 'The origin of life and the nature of the primitive gene', *J. Theor. Biol., 10,* 53–88 (1966).
7. Cairns-Smith, A.G., 'A case for an alien ancestry', *Proc. R. Soc., B189,* 249–274 (1975).
8. Cairns-Smith, A.G., *Genetic Takeover*, Cambridge University Press (1982).
9. Cairns-Smith, A.G., *Seven Clues to the Origins of Life*, Cambridge University Press (in press).
10. Miller, S.L., 'A production of amino acids under possible primitive earth conditions', *Science, 117,* 528–529 (1953).
11. Walker, J.C.G., 'Implications for atmospheric evolution of the inhomogeneous accretion model of the origin of the Earth', in Windley, B.F. (Editor), *The Early History of the Earth*, pp.537–546, Wiley, London.
12. Walker, J.C.G., *Evolution of the Atmosphere*, Macmillan, New York (1977).
13. Moorbath, S., 'The oldest rocks and the growth of continents', *Scient. Am., 236(3),* 92–104 (1977).
14. Allaart, J.H., 'The pre-3760 m.y. old supracrustal rocks of the Isua area, central west Greenland, and the associated occurrence of quartz-banded ironstone', in Windley, B.F. (Editor), *The Early History of the Earth*, pp.177–189, Wiley, London (1976).
15. Abelson, P.H., 'Chemical events on the primitive earth', *Proc. U.S. Nat. Acad. Sci., 55.,* 1365–1372 (1966).
16. Sillen, I.G., 'Oxidation state of the Earth's ocean and atmosphere 1. A model calculation on earlier states. The myth of the "prehiotic soup,"' *Arkiv Kemi, 24,* 431–456 (1965).
17. Rein, R., Nir, S., and Stamatiadou, M.N., 'Photochemical survival principal in molecular evolution', *J. Theor. Biol., 33,* 309–318 (1971).
18. Dickerson, R.E., 'Chemical evolution and the origin of life', *Scient. Am., 239*(3), 62–78 (1978).
19. Ferris, J.P., 'The chemistry of life's origin', *Chem. Engn. News, 42,* 22–35 (1984).
20. Kvenvolden, K.A., 'Natural evidence for chemical and early biological evolution', *Origins of life, 5,* 71–86 (1974).
21. Maynard Smith, J., *The Theory of Evolution*, 3rd Edn, p.287, Penguin, Harmondsworth (1975).
22. Fraser, R.D.B., 'Keratins', *Scient. Am., 221(2),* 86–96 (1969).
23. Pirie, N.W., in Bernal (Editor), *The Physical Basis of Life*, 71 (1951).
24. Millot, G., 'Clay', *Scient. Am., 240(4),* 76–84 (1979).
25. Theng, B.K.G., *The Chemistry of Clay-organic Reactions*, Hilger, Bristol (1974).

26. Cradwick, P.D.G., Farmer, V.C., Russell, J.D., Masson, C.R., Wada, K., and Yoshinaga, N., 'Imogolite, a hydrated aluminium silicate of tubular structure', *Nature (Phys. Sci.), 240,* 187–189 (1972).
27. Wada, S. and Wada, K., 'Density and structure of allophane', *Clay Miner., 12,* 289–298 (1977).
28. Bailey, S.W., 'Cation ordering and pseudosymmetry in layer silicates', *Am. Miner., 60,* 175–187 (1975).
29. Weiss, A., 'Replication and evolution in inorganic systems', *Angew Chem. Int. Edn., 20,* 850–860 (1981).
30. Keller, W.D., 'Classification of kaolins exemplified by their texture in scan electron micrographs', *Clays Clay Miner., 26,* 1–20 (1978).
31. Mansfield, C.F. and Bailey, S.W., 'Twin and pseudotwin intergrowths in kaolinite'. *Am. Miner. 57,* 411–425 (1972).
32. Fraser, J.T., *Time as conflict,* p.73. Birkhauser, Basel (1978).
33. Bull, A.T., 'Microbial growth', in Bull, A.T. *et al.* (Editors) *Companion to Biochemistry,* pp.415–442, Longman, London (1974).
34. Keller, W.D., and Hanson, R.F., 'Dissimilar fabrics by scan electron microscopy of sedimentary versus hydrothermal kaolins in Mexico', *Clays Clay Miner, 23,* 201–204 (1975).
35. Johnson, L.T., 'Clay minerals in Pennsylvania soils in relation to lithology of the parent rock and other factors—I', *Clays Clay Miner., 18,* 247–260 (1970).
36. Clemency, C. V., 'Simultaneous weathering of a granite gneiss, and an intrusive amphibolite dike near Sao Paulo, Brazil, and the origin of clay minerals', *Proc. Int. Clay Conf.,* Mexico, pp.15–22 (1976).
37. Lahav, N., White, D. & Chang, S., 'Peptide formation in the prebiotic era: Thermal condensation of glycine in fluctuating clay environments', *Science, 201,* 67–69 (1978).
38. Freund, F., and Wengler, H., 'Proton conductivity in hydroxides as related to transport phenomena in soils', Paper at the 6th Int. Clay conf., Oxford (1978).
39. Getoff, N., 'Reduction of carbon dioxide in aqueous solution under ultraviolet light. *Z. Naturforsch., 17B,* 87–90 (1962).
40. Getoff, N., 'Effect of ferrous ion on formation of oxalic acid from aqueous carbonic acid by ultraviolet irradiation', *Z. Naturforsch., 18B,* 169–170 (1963).
41. Brateterman, P.S., Cairns-Smith, A.G., Sloper, R.W., Truscott, T.G., and Craw, M., 'Photo-oxidation of Iron(II) in Water between pH 7.5 and 4.0', *J. Chem. Soc. Dalton Trans.,* 1441–1445 (1984).
42. Wey, R., and Siffert, B., 'Reactions de la silice monomoleculaire en solution avec les ions Al^{3+} et Mg^{2+}' in *Genese et synthese des argilles,* pp.11–23. R.S., Paris (1962).
43. Lipmann, F., 'Projecting backwards from the present state of evolution of biosynthesis', in Fox, S.W. (Editor), *The Origins of Prebiological Systems,* pp.259–280, Academic Press, New York (1965).
44. Hartman, H., 'Speculations on the origin and metabolism', *J. molec. Evolut., 4,* 359–370 (1975).
45. Schrauzer, G.N., and Guth, T.D., 'Hydrogen evolving systems. 1. The formation of H_2 from aqueous suspensions of $Fe(OH)_2$ and reactions with reducible substrates including molecular nitrogen', *J. Am. Chem. Soc., 98,* 3508. (1976).

Professor Raymond A. Lyttleton

Raymond Lyttleton is Professor Emeritus at the Institute of Astronomy, University of Cambridge. He describes himself as uneducated, but attended King Edward's School, Birmingham, and Cambridge University, where he graduated as a Wrangler, winning the Tyson Medal for Astronomy, and then proceeded to Princeton University, USA. During World War II, he served as Assistant to the Scientific Advisor to the Army Council, returning afterwards to Cambridge as a Lecturer. He is the author of seven books and over 70 major papers, is a Fellow of the Royal Society and President of the Milne Society.

He has attacked the problem of rotating gravitating liquids, and also shown how accretion of interstellar dust could account for the formation of comets. For these researches he was awarded the Gold Medal of the Royal Astronomical Society in 1959, and a Royal Medal of the Royal Society in 1965.

For the past two decades he has worked closely with the US Space Program, and this has concentrated his interest on the Solar System, particularly the structures of the terrestrial planets. These studies led to entirely new conclusions on the origin and development of the planets.

Journey to the centre of uncertainty

At first sight, this topic seemed rather formidable, and although I have spent most of my life at and near the centre of uncertainty, I could say little of ever actually making any journey *to* it, for I have always been there. (I felt rather like the Boston lady, who having surprised a friend by confessing she had never been on a train, easily dealt with it by saying, "Well of course not; you see I live here!")

What is science and its objective?

Scientific work is manifestly concerned with trying to understand something. If we feel we understand a matter, we do not bother much about it. But if this desk suddenly began to jump up and down, none of you could concentrate on what I am saying until it were "explained", and you felt you understood why it was happening. Then once the reason became clear, we could all relax and continue.

But, "How are we to *know* when we know?" When should we relax in a euphoric state of knowing that we have now come to understand something completely and with certainty? Actually the answer is 'Never quite ever'. Valuable as our attained 'state of knowing' may be, there may, or will always be, room for improvement. At one time, Newtonian dynamics was regarded as providing a complete understanding of the system of the world capable of telling us with sufficient effort all we needed to know or could know about the motions of the planets, and even the temporary difficulty retrieved by the discovery of Neptune in 1846 was destined to be very short-lived, for in 1859 LeVerrier showed that understanding of planetary motions, of Mercury in particular, was not as complete as had been believed, and many years were to go by before it became realized why Mercury did not conform to Newtonian theory.

Several noteworthy scientists have taken time off from 'doing science' to explain what they believe science is really about. Thus did Karl Pearson write his "Grammar of Science" with this object in view, and his main contention was that "The classification of *facts*, the recognition of their sequence and relative significance is *the function* of science." And he went on to add, "let us be *quite sure* that whenever we come across a conclusion in a scientific work (How can it be known *a priori* if it is this?) "that does not flow from the classification of facts, then we are dealing with *BAD* science."

But then along comes Eddington, senior wrangler, deep thinker and mathematician, theoretical and observational astronomer, all rolled into one, to warn us that no observation, nor any experimental result, can be relied upon or trusted *until it is understood theoretically*. A simple example of an observational *fact* is that the railway lines can actually be seen to meet in the distance. but if an impish engineer made them meet a mile or so away from us, the lines would still be seen to meet just as before. Of course, this example is simple to analyse and understand theoretically. But a far subtler instance led to a long and bitter controversy in the mid-nineteenth century. To explain this briefly, for most of his life Halley (1656–1742) was of opinion that the Moon was getting progressively out of position as based on pure Newtonian theory, but then in 1786 the great Laplace showed that a minute steady decrease was taking place in the eccentricity of the orbit of the Sun about the Earth–Moon system, and that when the effect of this on the lunar motion was allowed for it fully accounted for the observed acceleration of the Moon, as claimed by Halley, and it seemed to generally accepted satisfaction that observation and theory were brought into complete accord. Nearly 70 years were to go by before J.C. Adams, in 1853, challenged Laplace's calculations and showed that when these were carried to a much higher order of accuracy the discrepancy became restored, Halley's conclusion revived, and a long and bitter controversy ensued. On the one side, it was urged that since Laplace's result agreed with observation it must be correct, while on the other side several eminent mathematicians took up the problem to confirm the work of Adams, and it was Adams's conclusion that finally prevailed. Thus, it became proved that the Moon was in fact getting progressively out of position, as Halley had maintained, and the era of *tidal friction* ushered in, though the great significance of its consequences has still not been adequately appreciated by geophysicists despite the lapse of over a century. But here the relevance of this example is to demonstrate the soundness of Eddington's dictum. The observational result satisfied by the work of Laplace, though seemingly incontrovertible, was simply not properly understood theoretically. The theory was wrong, and the matter only became understood as a result of Adams's discovery of the correct theory. Thus a phenomenon of highest importance, tidal friction, failed to be recognized for a century through reliance upon an observation not fully understood theoretically.

But more perplexing still for the scientist, especially naive realists, long before Eddington's time, William Whewell (1794–1866) perceived a most profound property of the world around us that "*A mask of theory covers the face of nature*". All the apparently real affairs that we discern around us—buildings, the ground under our feet, the sky, and all the diverse associated phenomena, other people, indeed everything—are constructs of theory, but for the most part of theories that work so well and reliably that we hypostasize them, we consider them realities, and give them material and mechanical forms. However, in more complex realms, when the data are not understood theoretically, and cannot be got into any intelligible shape and order where now are poor Pearson's facts, facts to classify. Data, of a kind, he may have in embarrassing plenty, but not understood theoretically, and so no real *facts* to classify. If mere 'classification' were possible and all that was needed to 'do science', and if such classification could be carried out meaningfully without theory, then science would be easy to advance. But it does not turn out to be so, indeed very far from it.

Finding

We come now to the far more recondite question of where these all-important theories are to come from? Here, we are in a 'no man's land' with little or nothing to guide us. Raw data relating to a problem are probably available in abundance: in astronomy, we have smudges on photographic plates, for example, or numerous readings of instruments, and such like. But hypotheses and theory are needed that enable these raw data to be seen in some novel way so that they can become understood and transformed into *facts*. That such hypotheses have been found and theory built upon them, by inspired men of genius in science, is undeniable, and have led to the conquering of vast areas of phenomena, Newtonian gravitation and the laws of dynamics, for example, but still we may wonder how they were found, and found they must be for other realms if real progress is to be made. But there exists no formal standard procedure that will lead to the requisite hypotheses. Although the term is frequently used, there is no such process as *The* scientific method, or even a vaguely-definable scientific method.

In discussing some new realm, we are forced to introduce verbal terms for the purpose, but without really 'knowing what we are talking about'. Of necessity, these words may become established, which is why the verbal language of science is strewn with misnomers that may turn out to be quite inapposite but their use, nevertheless, continued on, and one has to learn that they do not mean what they seem to say and what they really do mean: 'seas' and 'mountains' on the Moon, to give an elementary instance, yet the phenomena to which they relate are still so termed. Then again, early attempts to classify the stars alphabetically proved quite unsatisfactory but have left us with the order O,B,A,F,G,K,M,R,N,S, indeliby inscribed in spectral theory. But no matter how forcefully and above all seemingly *logically* we are reasoning in using what is not recognized to be inappropriate terminology, nothing in the way of scientific progress may come of it. Logic itself is a kind of trimmed down distillate of reliable established theoretical knowledge, but may be of no help where new realms of thought are required, it may even be a handicap: witness the historic difficulty that was met with in 'reasoning' about transfinite numbers on the basis of properties of finite numbers. The great American inventor, Charles Franklin Kettering, showed that he knew this only too well from experience with his clever aphorism that, "Logic is an organized system of thought that enables you to be wrong with confidence!"

Knowing when we know

We have not yet answered how we are to know when we know. That is, having arrived at a hypothesis and having constructed a theory of some phenomena by means of it, how are we to judge if it is a worthwhile theory, or whether it is at any rate closely correct? The great Poincaré, mathematician and scientist, understood this matter thoroughly, and to start with ruled out all merely descriptive verbal accounts of phenomena as not amounting to scientific theory at all. In explaining the true role of scientific theories, he first laid down the principle that "Physical theories should not introduce as many (or more) arbitary constants, or verbal assumptions, as there are phenomena to be explained". Then he added that, "They should establish connections between different experimental results (and observations), and *above all*, they should enable successful predictions to be made." This last requirement is the ultimate answer to our question. It is at this stage, when theory is able to make predictions that subsequently turn out to be correct, that we can say that the phenomenon is understood, that we *know* that we know about the realm concerned.

We all realize this instinctively in day-to-day affairs. There are so-called memory-men who can get up on stage and tell you the name of the winner of the Derby in any year, right back to the year dot (1780 in England). The audience may marvel at such a feat, though in content it really is no more than an impressive 'descriptive' account of the nature of a series of names, but one would really like to meet the man who could tell you next year's winner, not by some crystal-ball gazing, but by means of some reliable scientific theory. Alas, no such man exists or ever will. But from time to time a few such geniuses do arise in science who can by some inner light see how to go about striking order into the data of some hitherto inexplicable phenomena, and not only go on to build a theory that shows the relationship between the phenomena, but a theory that predicts how the natural system will behave in given circumstances *in the future*. These are the great scientists, with Newton heading the list, and so far-reaching in application were his simple laws of dynamics and gravitation that even to this day they are still by no means fully worked out, and there are still minor mysteries. But it can be left to lesser mortals, with the theory already available and established by its vast accumulation of successful predictions for complete reliance to be accorded to it, to tackle the numerous extant problems to which the theory almost certainly applies and would create a sensation if it were found not to do so.

Correct attitude

We now consider what attitude a scientist should adopt towards ideas, hypotheses, and theories, especially those of a novel kind, but also only to a slightly less extent towards hypotheses and theories already well established. It seems to me that most people, including many scientists, crave for completely certain knowledge, for certainty of belief or of disbelief, in regard to ideas, yet this is not a worthwhile or valuable attitude for a scientist, or anyone, to adopt. I think the correct attitude can be illustrated geometrically very simply (Figure 1) by means of a small bead that one can move to the right or left on a finite length of a horizontal piece of wire.

Suppose the left-hand end of the wire, marked 0, corresponds to an attitude of complete unqualified *disbelief* in some particular hypothesis, and the right-hand end, 1, corresponds to an attitude of absolute unqualified *certainty* of belief in to idea. Just where one places the bead on the wire can obviously represent an indication of one's individual personal attitude to the hypothesis under discussion. I believe, and we are talking subjectively, the principle actually to be practised that I would urge on any intending scientist in regard to any and every hypothesis and theory should be:

Never let your bead ever quite reach position 0 or 1

This is quite possible, for however close to the end one may have set it, there are still an infinite number of points to move the bead to in either direction in the light of new facts or new arguments or whatever. There is no need to keep your bead at a definite fixed point on the wire, *but* only ever move your bead as a result of your own independent revised judgment of the validity of the hypothesis prompted by new scientific facts, or by new arguments, or by anything new that may be relevant. If this revision increases your confidence in the hypothesis, then move your bead towards 1, by what amount is for you alone to judge. On the other hand, if the revision lessens your confidence, then move it towards 0.

All this is quite possible, for however close to the end 0 or 1 you may have set your bead, there are still an infinite number of points to move it to without the bead going off the wire. But never-never move it purely on the conclusions or opinions of others without first critically examining the basis of their belief and deciding for yourself its scientific validity. And perhaps the most important precept of all: *never be influenced by mere*

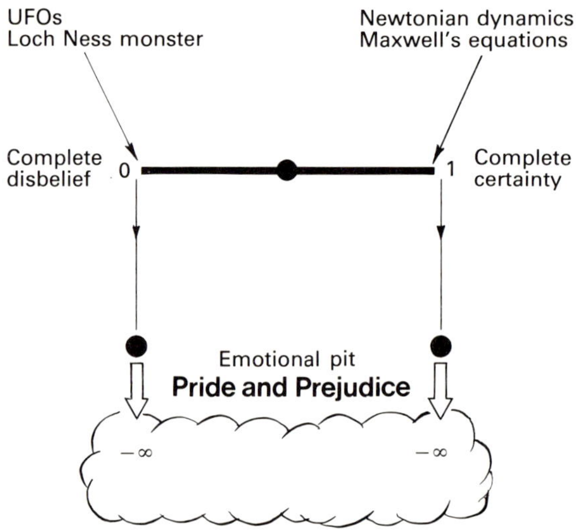

Figure 1 The attitude that should be adopted by scientists, shown by analogy with a bead on a wire.

numerical strength of believers or disbelievers. (It is often claimed that 50 million Frenchmen can't be wrong, but they can, and a great many of them may have paid for their beliefs with their lives!) Yet many people regard the opinions of others, especially if they occupy positions of high rank, as providing a sound basis for their own views. It is, of course, a form of sloth enabling them to spare themselves the hard work of properly forming their own conclusion, and instead just take up a ready-made one issued by the establishment free of charge, just as one might wear a ready-made suit however ill-fitting. This was the point that Einstein had in mind when he said, 'If most of us are ashamed of shabby clothes and shoddy furniture, let us be more ashamed of shabby ideas and shoddy philosophies.' Then again, well aware of this widespread defect of judgment of many people, as late as 1930 a conspiratorial group of supposed scientists in Germany brought out a book entitled "A Hundred Professors Prove Einstein Wrong", with the transparent object of persuading non-thinking 'scientists' and the general public to move their beads, that is, if they were still on the wire. On hearing of the publication, Einstein simply said, "Well, *one* would be enough!" Of course, a true scientist made aware of such an incident would if anything be more likely to move his relativity-bead in the opposite direction to that hoped for by the conspiring professors.

A further advantage of this bead-practice, is that by keeping the bead always on the wire, one can lead a calm and tranquil life, even in the midst of heated controversy by others (with their beads off the wire), and not allow such conflict to impair one's judgment. This is not simply a question of 'not taking sides' in order to keep out of the fray, or of 'taking sides' without giving equal study to both sides but just 'sitting on the fence', for as C.S. Lewis so wisely pointed out, "There are two" (or more) "sides to every question—until the truth is known." Nor will the bead-method cause you any great emotional upset if some new fact or line of argument comes along that requires the relevant bead to be moved. General relativity did this in relation to Newtonian mechanics, and many whose beads in regard to the latter had long since been well off the certainty-end of the wire were much perturbed and hostile to what seems a very natural and welcome development of gravitational theory. Eddington for one took the revised theory in his stride (he was still a young man), and wrote a mathematical treatise of classical erudition on the subject. Then again, which may be a source of satisfaction, with one's beads safely on the wire, one will never have to 'eat one's words', but at most suitably modify them without undergoing pain.

But if one once goes completely overboard for an idea—the bead goes right off the end of the wire—then one's attitude falls into an irrational realm, and it may prove difficult ever to extricate it and get the bead back up onto the wire again, if indeed the owner of the bead, having so committed himself to the idea, should seriously wish to get it back on the wire, even if he could do this of his own volition, and even though it would be to his own scientific advantage. In such a case, it may even be quite unwise to attempt to interfere with another person's bead, as he will probably be very proud and self-satisfied with it remaining stuck in a morass of deep irrationality: any suggestion that he might well move his bead back up onto the wire may 'rouse the passions', as the French put it. If one should attempt it by means of reason (and not by authority), much tact may be required, but in many cases the bead may have fallen so far into the depths that it would be madness even to try, and in some cases positively physically dangerous.

A lecture by Sir Peter Medawar some years ago was entitled "Advice to a young scientist", and one of his most emphatic pieces of advice was *never fall in love with your hypothesis*, by which he meant, although he did not use the bead-and-wire analogy, 'Never let your bead fall off the end of the wire'. But we see 'scientists' so enthusiastic about their idea as to be driven to shift after shift of extravagant adjustments and additional *ad hoc* assumptions to try at any cost to *save* the precious well-beloved hypothesis, thereby getting deeper into a mire of confusion, when what the situation is really implying is that the hypothesis is simply plain wrong. In a subsequent write-up of his talk, Medawar put the precept far more explicitly by stating "I cannot give any scientist of any age (young or old) better advice than this: *the intensity of conviction that a hypothesis is true has no bearing on whether it is true.*"

The Gold effect

The danger in accepting a scientific theory can only be avoided by taking nothing for granted on the opinion of others and making it a rule always to rely on one's own independent scientific judgment, even to the extent of checking the arithmetic and mathematical steps of a paper for their accuracy before accepting the conclusions. This may be made more difficult if some inadequately-defined computer program has been utilized but not allowed to be open to inspection. In such cases, extra special care should be taken in accepting any results claimed to depend on the program.

Professor T. Gold has stressed the importance of relying on one's own scientific judgment in such cases, by drawing attention to a remarkable multiplicative effect, appreciated by few if any, that must almost always enter into the mechanism of how beliefs may come to be entertained and then gradualy go on to be widely regarded as well established when there may be no scientific basis for them whatever. To explain, let it be supposed in regard to an idea that a number M of people are interested in that idea. Diagrammatically, let the idea be represented by a point 0 on a straight line, and the degree of non-acceptance, of adverse criticism, or of disbelief in the idea as proportional to the distance x from 0 along the line in either direction for simplicity. The attitude of any particular one of the M individuals can be used to classify him in a small range from x to $x + dx$ of the line. Unqualified belief in the idea will place that person's representative point at 0, and strong disbelief far out towards $+\infty$ or towards $-\infty$. Next let it be supposed, as is likely to be the case, that a slight preponderance of the M people have arrived at a state of near-belief in the idea, so that there occurs in the distribution of the M people with x a weak maximum at 0. A suitable function adequately representing this for present purposes would be the gaussian exponential function $\exp(-x^2)$, with the scale of x sufficiently large to give a weak maximum at $x = 0$. Then the distribution of people by number in relation to the idea would be

$$M(\pi)^{-1/2} \exp(-x^2)\, dx$$

the numerical factor $(\pi)^{-1/2}$ securing that the number of people from $-\infty$ to $+\infty$ is equal to M. There is nothing special to the argument (to be given) about this particular function, though it is convenient numerically for plotting curves, and any similar function or distribution-curve initially exhibiting the slight peak about $x = 0$ would serve equally well and lead to results exactly similar to those that will be demonstrated.

Now if this represents the situation at a given stage, suppose some action is agreed upon in regard to the idea. For example, it may be decided to hold a meeting to discuss the pros and cons of the idea. And let it further be assumed that every effort is made to give all concerned equal opportunity to attend and to state their views and arguments. Perhaps attendance might need to be restricted, and names drawn out of a hat to effect this, but the attitudes and distribution of those attending would, nevertheless, automatically reflect the existing distribution. Thus there will be present more people favouring the idea than against it in equal ranges of x, and when all is said and done and the meeting over, if the people are influenced by other than strictly scientific considerations, (and the whole purpose of meetings and writings is presumably to influence other's opinions), the distribution in regard to the idea will now be in proportion with $\exp(-x^2)$ *multipled by* $\exp(-x^2)$, that is to $\exp(-2x^2)$. The functions or curves will not be additive, for that would leave the distribution still proportional to $\exp(-x^2)$, and therefore unchanged: which would mean that the meeting had had no effect on the general attitude to the idea.

A subsequent step might be to appoint a 'representative' committee to get out a collection of papers to propagate and foster further interest in the idea, and for this a list of contributors must be drawn from the M people. Again, there might be too many to invite all concerned, but even if the committee and authors were chosen entirely at random, each of the selection will have a distribution roughly but strongly proportional to the existing distribution, $\exp(-2x^2)$, and the total resulting articles will appear to show an increased consensus of agreement and acceptance of the original idea but now proportional to $\exp(-4x^2)$. And so it will go on, each and every step taken will be directly influenced by the currently existing distribution, and lead to greater and greater concentration of the peak of certainty at 0. For instance, with papers submitted to a journal relying on a refereeing procedure, the selection of referees will obviously be influenced by the selectors and by the ready field from which referees are available. Without any ulterior motive, unless all concerned are from the very start influenced simply and soley by correct scientific considerations (which would require M to be very small indeed!), after n such operations, the distribution of attitudes will have become proportional to something like $\exp(-n^2x^2)$. So that numbers total up to M (which may now be different from the initially interested M), the distribution function will have become

$$M(\pi)^{-1/2}\, n \exp(-n^2x^2)\, dx$$

and the peak at the origin 0 will have become n-times as strong as at first. As time goes on and n gets larger, the distribution becomes concentrated more and more round the idea having the initial slight advantage from whatever cause, there are fewer and fewer outliers, and an appearance of

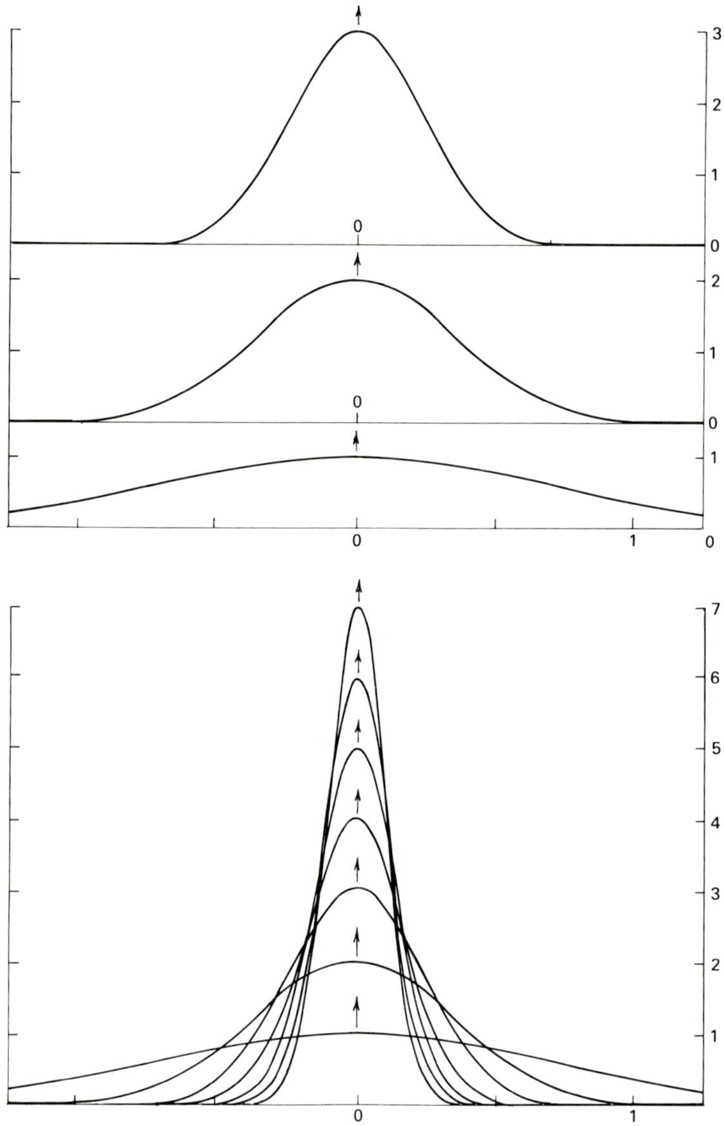

Figure 2 Geometrical illustration of the Gold Effect. The upper shows separate curves of $n \exp(-n^2x^2)$ for n=1,2, and 3, the total area under each curve being the same and corresponding to the number of people concerned. The lower shows superimposed seven successive equal-area curves, demonstrating the tendency for concentration around the origin 0 as n steadily increases. The ultimate state would be a so-called delta-function, attaining to the infinite at 0 with a host of true-believers and zero non-believers. This is the 'ideal' always sought by obscurantists bent on a search for certainty and the annihilation of all dissent. (The exponential function has been used solely for numerical convenience in drawing the curves. Any function or curve possessing initially a weak maximum at 0, or a series of such similar functions would lead to the concentration around 0.)

almost complete certainty of the correctness of the idea can come about in the minds of most people purely through this multiplicative numerical effect. Of course, in practice it does not work with the uniform smoothness assumed here to clarify the discussion, but the tendency for an initial weak maximum to become enhanced more and more of its own accord will always be present, unless none other than purely scientific considerations are ever brought to bear by all concerned, which as aforesaid would be completely impossible to suppose.

This then is what has been termed the Gold Effect, after its discoverer. In essence it is that even if only random moves and selections are made in regard to an idea initially supported by a slight majority, then through lack of rigorous scientific thinking and allowing other considerations to be allowed to have influence, there will develop an increasing concentration of people believing that the idea is correct, when, in fact, it may be no more than an arithmetically brought-about illusion of truth, a simulacrum without scientific basis, a kind of lying truth. Figure 2 illustrates the progression of the Gold Effect by means of a succession of Gaussian curves of increase (negative) index, as explained above. The area under each curve is the same, and in suitable units would be equal to the total number of people concerned at any time.

The operation of the effect has been described in terms of its simply taking its natural course propelled purely by existing lack of adequate analytical faculty, but it must be recognized that, in reality, other influences are brought to bear that cause the effect to proceed more swiftly and certainly than if left to occur at its own rate. Few of those associated with science are supermen, free from all weakness and able to remain above the conflict; indeed, there is no reason to suppose that scientists as a class, whatever their opinions of themselves, have one and all been born free of any modicum of human frailty making them morally superior as a whole to any other group of people. Men in science can often be recognized to be unduly proud of their work, usually in inverse proportion to its worth, and with ordinary unthinking mortals as has been seen become so enamoured of their ideas as to be incapable of being adequately critical of them. Then again, although theories may be propagated sedulously that have no scientific validity, most people by far seem incapable of perceiving this through lack of analytical capacity and of any understanding of the nature of knowledge. Witness the enormous popularity and widespread acceptance as fact of the works by such as Velikovsky, and the claims made by charlatans to possession of occult powers.

Examples

The foregoing describes the basic principles of operation of the Gold Effect. But we can now go further and give specific examples of how widely believed, but fallacious or erroneous, conclusions have come to be regarded as providing fundamental established results ruling a whole subject. We begin with modern geophysics.

First instance is provided by the widely held belief that the moment-of-inertia of the Earth has remained constant from the time of its formation. This unquestioned view seems first to have been foisted on the scientific world by Jeffreys[1] in 1920 as the main error in an incorrect attempt to determine the lunar and solar tidal couples from the recently available ancient-eclipse data, and seems to have been generally accepted without the smallest question just on the sole basis of Jeffreys' 'authority' for over half a century. Not until 1976 was the error pointed out publicly that, with C and ω denoting the moment-of-inertia and angular velocity, Jeffreys assumed that the rate of change of the angular momentum (as a result of the action of the lunar and solar couple) was $C\mathrm{d}\omega/\mathrm{d}t$, whereas it should be $\mathrm{d}(C\omega)/\mathrm{d}t$. It turns out that if the form $C\dot{\omega}$ is adopted, no physically possible solution exists of the equations that emerge for the couples, the solar couple coming out large and *negative*. Had a solitary reader troubled to check the elementary arithmetic of Jeffreys, he must have discovered the physically impossible result that the value of the solar couple came out negative, a result itself cleverly concealed by Jeffreys in this 1920 paper. In fact, when the correct form, $\mathrm{d}(C\omega)/\mathrm{d}t$ = couples, is used $\dot{C}$ is importantly negative, and its far-reaching consequences completely change the whole theory of the evolution of the Earth, including for the first time an adequate theory of mountain-building. Even after its being pointed out in 1976, the result was ignored and tacit acceptance of a constant C persisted. Not until 1982 after four years effort was a publisher prepared to fly in the face of general accepted geophysical theory, and produce the first textbook giving the correct mathematical equations for the dynamical analysis of the lunar and solar couples from the known ancient-eclipse results, while hostile reviewers continued to describe already existing books, replete with entirely erroneous dynamical theory, as classics of geophysical literature. The situation is wholly an example of the Gold Effect going to the full limit of its uncritical operation, and has resulted in volumes of slothful papers extolling work as 'classical' when it is nothing more than a series of errors.

A second example from geophysics is the widespread notion that the dense outer core of the Earth consists of liquid iron. In 1906, it was discovered by the seismologist R.D. Oldham that a large central region of the Earth, extending out some 55 per cent of the radius, behaved as a liquid in not transmitting secondary(S)-waves, only primary(P)-waves, but no conclusion as to its composition could be arrived at, since this is not revealed by seismology. More than 20 years later, in 1929, Oldham's discovery was deemed worthy of mention by Jeffreys whose account of Oldham's work ends with the unexceptionable statement that, "this at once gives a strong suggestion that the Earth has a large liquid core in the middle," which, of course, was Oldham's unembellished result. But only a little later in the account by Jeffreys, the phenomenon is broadened conjecturally to the form "that the Earth has a dense core", and for no particular adduced reason is added, "*probably* composed of liquid iron." But after 10 pages devoted to other questions it is stated outright that "The iron core hardly affects these considerations." And from then on the core is always referred to unequivocally as 'the iron core', till finally much later, a discussion of the figure of the Earth ends categorically with the statement that "the rock shell is of one material from a depth of 30 km to the great (sharp) discontinuity, and that the central core is of *liquid iron*." Thus was 'the iron core' born in 1929, and is still going strong, driven deeper and deeper into the dark recesses of the minds of geophysicists (and many others) as one of the central tenets of the subject by sheer iteration in accordance with the Gold Effect, till now more than half a century later it has come to be regarded as an established dearly-loved fact accepted without question by all sane respectable geoscientists.

When the uncompressed density of the outer core was shown from seismic results to be far less than that of iron, the difficulty was immediately overcome by watering-down with addition of suitable amounts of oxygen *or* sulphur *or* silicon, exactly which is not clear, though just how the requisite reduction in density was established in the laboratory, or theoretically, for pressures well in excess of a million atmospheres and temperatures of a few thousand degrees is difficult to understand unless the standard geophysical means of verbal assertion is accepted as constituting proof. But even under ordinary conditions, a compound of a metal with a gas may not necessarily have lower density than the original metal.

The 'iron core' has so deeply entered the hearts and souls of the geophysical establishment that attempts to 'save' it have been made by comparison with shock-wave experiments, but the conclusions are quite fatuous. In these experiments, the velocities of propagation of shock-waves, $(\delta P/\delta \varsigma)^{1/2}$, through selected metals are compared with those of the seismic waves in the mantle and core which of course change with depth. Diagrams making this comparison were first published by Birch in 1961 and then again in 1963. In the first, the seismic curves of velocity against density for the two Bullen models of the core never come near two closely similar shock-wave curves available for iron. But in the 1963 version of the diagram, Bullen's seismic curves for the core, which meanwhile had gone unchanged since 1961, have mysteriously got moved over bodily to the right so that one of Bullen's curves now just intersects the shock-wave curves for iron, but even then at an angle of about 30° (Figure 3).

The reader is left to form his own opinion of the lateral shifting of the

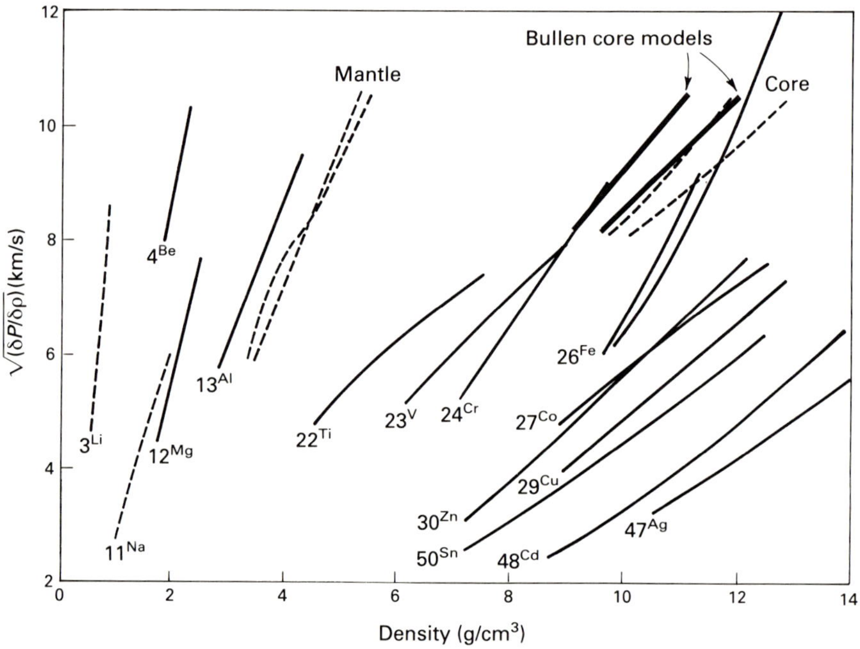

Figure 3 Comparison of seismic and shock-wave velocities.

Bullen-curves and to decide to what extent a 30° angular fit can be regarded as the close fit that is claimed. These, however, form the basis of the alleged shock-wave 'proof' that the core can be identified with iron. But if attention is turned to the mantle, Figure 3 shows that the seismic velocities therein are far closer to the shock-wave velocities for aluminium than they are for iron in the core. If the 'arguments' for the latter establish that the core is of iron, they show even more strongly that the mantle is aluminium, which remarkable conclusion passes unmentioned.

The opinion of an independent referee was sought on the above work, a colleague renowned to be of highest competence in this field. He reported back as follows: "I agree with your criticism of the comparison, namely that in the seismic context an infinitesimal wave moves through material already *under high pressure* and a temperature of a few thousand degrees, whereas in shock-wave experiments a *severe shock* moves into material initially *unstressed* and at laboratory temperature. The treatment-history to be accorded to a small volume-element of material is *totally different* in the two cases, and the resulting thermodynamic states must be so different that no valid comparison is possible." And he added, "in my view Birch's arguments are complete nonsense." So much for the shock-wave demonstration that the core is of iron, but it is still widely quoted as providing convincing 'proof'.

An indirect third line of defence of 'the iron core' was taken up by F.C. Frank in a paper[2] offering 'thermodynamic proof' that the Ramsey collapse is physically impossible. Although my calculated mass of the initial liquid core after collapse is 3.708×10^{26}g, Frank prefers to consider the melting of a central sphere of mass only 10^{26}g, and for his purposes needs to calculate the pressure at the surface of this sphere, which he finds to be 1.2×10^{10} dyn/cm^2 less than the central pressure. In fact, the correct value of the difference is about 1.1×10^{11} dyn/cm^2, a mere factor of about 10 in error, but the value is crucial to his whole 'argument'. When this correct figure is used in Frank's arithmetic, the final conclusion is completely reversed, and his theory (if valid, in principle) proves that there is substantially more than sufficient energy available for the sudden liquefaction and collapse.

Oblivious of the numerous large errors that would result from that noted above, Frank admits that slow growth of a core might occur in 10^9 years, but never as a *sudden* event. He prefaces this with the unqualified claim that "of the few things that can be said with certainty about the nature and origin of the core one is that it did not form in the way described by Lyttleton." However, he makes plain that nothing in his paper impugns the brilliant work of Ramsey, but then concludes by giving references just to Ramsey's first two papers. But Ramsey wrote at least *five* major papers and some minor notes developing his theory, and had Frank read these before extolling Ramsey's work he would have found Ramsey concluding in his third and main paper that, "The time taken for the collapse is comparable with the time taken for a free particle at the surface of the planet to fall through 100 km under gravity, that is a matter of minutes or at most hours"—rather less than 10^9 years!—and he goes on that "in this short space of time, energy of 10^{36} or 10^{37} ergs is liberated in the collapse (the more accurate calculation by the writer is 5.94×10^{37} erg) and is supplied by the gravitational field." (See ref. 3.)

However, no retraction of Frank's work has ever been made nor have the Society originally 'responsible' for its publication published any paper pointing out the errors of Frank's paper and its obvious conflict with Ramsey's well-established theory. Thus does the subject of geophysics steadily become more deeply entrenched in its crude but well-loved misconceptions.

Cometary crudities

A second illustration of the Gold Effect is a vast, but completely baseless, astronomical set of inventions, replete with inadmissible assumptions and invalid mathematical and physical argument that has come to be termed 'modern comet theory'. It consists of two main planks: first, the so-called Oort-shell of some 10^{11} comets, all at too great a distance from the Sun to be detectable telescopically; and second, the icy-conglomerate nucleus model for the physical structure of comets, a nucleus of kilometric size and thus far below the resolving-power of any telescope. These features have enabled the theories based on them to escape any test by observations and so to survive by mere asseveration. It will be convenient to describe their deficiencies separately since the two notions have little in common except their complete erroneousness.

The Oort-shell. All long-period comets have keplerian orbits when near the Sun inevitably with large values of the major-axis, $2a$, and values of the eccentricity very near 1, with about one in eight exceeding unity—the so-called 'hyperbolic comets'. But to know the greatest distance from which a long-period comet originally came in, it is necessary to compute the orbit backwards in time allowing for planetary attractions until these become negligible and the comet is moving in purely elliptic motion about the barycentre of the Solar System. The long period comets for which such calculations have been made number about 50, some 10 per cent of the whole, but considerably fewer were available at the time of Oort's first publications. At great distance from the Sun, $2a$, of course, remains large, but the osculating value of the eccentricity e always takes a value less than unity. Using such values of a, Oort plotted a histogram of numbers of comets against equal intervals of $1/a$ of 5×10^{-5} AU^{-1}, and made the 'great discovery' that there was a strong peak in the diagram close to $1/a = +0$. The argument then went on to infer that the diagram implied that there were about 10^{11} comets moving about the Sun but confined to motion within a vast spherical shell of inner radius about 30,000 AU and outer radius about 100,000 AU. But the argument is absurd, since making the plot against $1/a$ has the effect of inverting arbitrarily large volumes of space into minute ranges of the abscissa near $1/a = 0$, and a strong peak duly appears near $1/a = 0$, but indicative of nothing more than that all comets made use of have large a-values. If a plot of numbers of comets against a itself, a quantity directly representative of volume, is made, there is not the slightest sign of any congestion of cometary aphelia in the range 30,000–100,000 AU or anywhere else. In defence of the $1/a$ method of making the plot, it has been urged that $1/a$ is proportional to the energy and 'therefore' the natural parameter to use. But $a^{3/2}$ is proportional to the period (and to the square root of volume), and if it is used there is again no sign whatever of any heavily populated shell at distances ranging out to 10^5 AU.

This large number of order 10^{11} comets in the shell is deemed necessary because this then allegedly allows stars passing near the shell to deflect comets through large angles into the inner reaches of the Solar System with sufficient frequency to account for the observed appearance of long-period comets, of some 5–10 per year. But it seems to have been overlooked that there are long-period comets with periods of order 10^4 years and less, and the stars that deflected them in would still be adjacent by stellar standards and therefore brilliantly visible, unless the further assumption is made that these stars were all dark and invisible. Moreover, a comet that was at distance D when so deflected by a star would not reach the Sun as if coming in from that distance, for its energy would be markedly changed and with it the value of $1/a$, and thus the dimensions of the shell could not be inferred from the calculated greatest distances.

As to the origin of the shell itself, here the theory really excels itself in its fatuity, for the claim is that they are asteroids (sic) ejected by passing stars from the inner regions of the Solar System out to the distant territory of the shell where *mirabile dictu* their paths are then rounded up into the shell again by passing stars for the necessary few to be deflected back in again by passing stars to become visible as long-period comets, miraculously transformed from asteroids. Here we must have a *deus ex machina* par excellence. It need hardly be said that no dynamical investigation of the claims has been attempted by the proponents who rely entirely on verbal assertion that the processes would work in just the required way, but there is nothing in the whole of celestial mechanics justifying such claims. All these criticisms and more besides have been published on numerous occasions but without the smallest effect, and a whole series of worthless papers have appeared claiming to prove all sorts of weird and wonderful properties for this obviously non-existent shell.

This $1/a$ argument calls to mind the famous spoof-paper on "The mathematical theory of big-game hunting." For simplicity the problem is reduced to that of catching a lion in the Sahara Desert by mathematical means, and one geometrical method proposed was for the hunter to set up in the desert a locked spherical cage with himself *inside*, and, of course, the lion *outside*. The hunter then inverts the cage into itself, and he is now outside, while inside is the hapless lion duly captured. But notice he is by no means alone but has the company of all the long-period comets—10^{11} of them!—and a great many other items to add to his discomfort. Yet the Oort-shell, through the ceaseless repetitive working of the Gold Effect has become widely regarded as a firmly established triumph of 'modern cometary theory' when in fact, it is a piece of trash heralded as one of the corner-stones of cometary 'science'.

The icy-conglomerate nucleus. Considerable prominence began to be given about 1950 to the idea that at the heart of the coma of any and every comet is a solid core of dimensions of the order of a kilometre, and that it is this nucleus that is the fount and origin of all the great variety of cometary manifestations. The idea has been put forward a number of times but never taken seriously in the face of actual cometary observations, whereas since its latter-day proposal it has provided opportunity for an increasingly great outpour of papers based on assumed *ad hoc* properties of this nucleus before it has ever been established that it has any basis in observation or in theory. In fact, the idea is ruled out at once by Occam's principle (AD 1349) which forbids the introduction of unnecessary hypotheses and assumptions. Since a length of 1 km at distance 1 AU would subtend an angle less than 0.002 arc seconds there can be no telescopic evidence for this micro-nucleus, and there is certainly no *a priori* theoretical evidence. Actual comets are seen as vast diffuse and irregular patches of light 10^5 km or more in dimensions.

There is no doubt that cometary heads show intense internal activity when near to perihelion, and in order that something should happen to this nucleus as it approaches the Sun, it is assumed to be constituted of dust and volatile ices, and for different comets, which have different perihelion distances, a series of icy substances are postulated that in each case vaporize at just the right distance from the Sun, and it is assumed that vaporization leads to a huge coma ($\sim 10^5$ km) every time the comet returns to the Sun, and capable of giving rise to a gas and/or dust tail. It is assumed that dust particles of requisite size are already embedded in the ices, but how they come to be there is as unanswered as how the ices form, but are assumed to be released when the latter melts and gasifies and thereby blowing the dust particles outwards from the nucleus to form a vast gaseous and dusty coma, but actually observation shows that the coma contracts as it approaches the Sun roughly in proportion with its distance. Anything that it is wished to come out during this activity is simply put in beforehand by special assumption. The seeming difficulty that comet Humason showed a fine tail at more than 3 AU from the Sun, whereas Ikeya-Seki got well inside the orbit of Mercury and even to perihelion before really extensive tail-development was observed, is dealt with by assuming that the former comet contains ices that melt and evaporate at the low temperatures prevailing far outside the orbit of Mars, while Ikeya-Seki is made of ices that require near-solar temperatures before melting and gasifying. Indeed at the stage when Ikeya-Seki seemed not to be behaving as an icy-nucleus should, it was claimed that the comet was 'worn out' (no doubt by the burden of assumptions it was carrying), but the claim was quickly dropped after it passed perihelion and became a brilliant 'daylight comet' with an impressive naked-eye tail. Similarly, Comet Stearns, which was followed out to about 11.5 AU, where 1 km would subtend less than 0.0002 arc seconds at Earth is 'explained' to be visible by the assumption that it consists of clathrates capable of producing an observable gaseous coma when at a temperature as low as about 100 K.

Despite more than 30 years sedulous propaganda for the nucleus 'theory', no remotely satisfactory process has been concocted as to how and where the 10^{11} snowballs inlaid with infinitesimal dust particles (capable of being repelled by solar radiation) ever got put together, and why they should attain to sizes of kilometric order only and then stop with nothing larger, whereas solid planetary bodies, whose mode of formation is understood, range from less than 1 km diameter to more than 10^4 km. The most impressive(?) notion to be managed to date is the well-nigh unintelligible assertion that "comets originate by the aggregation of cosmically abundant condensates from typical interstellar or stellar gases." The mere statement of such as 'scientific theory' for their origin seems criticism enough of its prolix absurdity. The remarkable orbital properties of comets are not even remotely explicable by any of the numerous *ad hoc* assumptions of 'modern' comet theory. That the long-period orbits have a mean-plane almost coincident with the galactic plane, which is far removed from the planetary plane is simply ignored, and in defiance of this well-established result it has even been claimed that the comets represent objects left over un-coopted at the time of formation of the planets.

The loss of mass through tail-production and formation of meteor-streams of an 'average' comet has been claimed on the basis of the nucleus-structure to range from 0.1 to 0.5 per cent of its whole mass at each return. Such mass-loss even if nearly correct would raise no immediate future difficulty, it seems to dispose of the theory that if such loss applied to Comet Halley, for example, which not only still produces an impressive tail (1910) but also has an associated meteor-stream, both implying loss of mass, then proceeding backwards in time the mass would have been comparable with that of the Sun in less than a million years. That some comets have been seen to divide into two or more main parts is taken in its stride on the icy-nucleus theory by simply making the further assumption that the disruption is by radioactive heating of the nucleus. But any admissible content of long-lived radioactive elements could raise the temperature of a sphere of radius of order 1 km by only a fraction of a degree above the ambient temperature from solar radiation, and merely simulate minutely closer proximity to the Sun rather than promote violent large-scale break-up. The assumption would also raise the question why only a mere few comets contain radioactive materials effective in the manner claimed while most apparently do not. Tidal action of the Sun has also been assumed as an agency for splitting comets, but for the Sun-grazers apart a micro-nucleus would have far too high a density. Tidal disruption has recently been assumed as the explanation of the several nucleuses shown at different times by Comet West, but with its perihelion distance of about 0.2 AU, the mean-density of the nucleus would need to have been less than 10^{-5} g cm^{-3} for solar tides to have any effect, while most ices would have density of order unity.

One feature above all that would be expected to hold for a kilometre-sized nucleus is that, as does an asteroid, it would describe a purely dynamical path but, in fact, many short-period comets do not do so and exhibit marked departures from strictly gravitational orbits, some at times showing accelerations in longitude and others retardations. On the nucleus theory, these seeming anomalies are readily 'explained' in the standard way by introducing further elaborate assumptions to the effect that such comets are subject to non-gravitational forces represented by fairly complex formulae that contain arbitrary constants different for different comets, forces which it transpires change in magnitude for the same comet from one return to the next, sometimes even as to sign. The forces are ascribed by still further assumption to asymmetrical jet-propulsion as a result of rotation of the nucleus which displaces the direction of thrust resulting from evaporating gases, a rotation so assigned as to give whatever acceleration is found to be required for that particular return of any specific comet. The introduction of arbitrary parameters, to be solved for in fitting the observations is found to allow reduction in these orbital discrepancies, as, of course, is no more than would be expected mathematically. But this does in no way necessitate or imply the presence of a kilometric-nucleus, as the procedure of introducing *ad hoc* forces could, with at least equal validity, be applied to the observations without raising the question of the physical structure of comets at all.

It is not realized by these would-be theorists that the need for special assumption after special assumption in order to 'save' the theory is simply nothing more than indication after indication that the original hypothesis is incorrect, whereas the long list of assumptions are perversely and proudly regarded by them as 'discoveries' of actual properties of comets. The resulting fairy-story theory provides a typical example of what Langmuir has termed Pathological Science, or "The science of things that aren't so." Astronomical literature and its readers could have been relieved of the enormous bulk of publication devoted to propagating this hypothesis of a non-existent nucleus if regard could have been paid to the true nature of scientific theory as enunciated by Poincaré.

References

1. Jeffreys, H., "The chief cause of lunar acceleration", *Mon. Not. R. astr. Soc., 80,* 309–317 (1920).
2. Frank, F.C., *Geophys. J. R. astr. Soc., 23,* 461 (1971).
3. Ramsey, W.H., *Mon. Not. R. astr. Soc., 110,* 334 (1950).

Dr. Richard Voss

Richard Voss is a research scientist at the IBM Thomas J. Watson Research Center. His first degree in physics was achieved at MIT in 1970, followed by a PhD from the University of California, Berkeley. He is a member of the American Physical Society, and has published over 40 papers on a wide variety of topics.

His major research interests in physics include work on Josephson Junction devices, the foundations of quantum mechanics, macroscopic quantum tunnelling, and associated work on superconducting phase transitions and weakly coupled Josephson Junctions. In the area of Fractals, his work includes investigations in time and space domains, represented by statistical music analysis and stochastic composition, random geometry and growth, and percolation studies. His use of computer graphics for Fractal imagery has produced results that are both richly informative, and artistic reflections of the physical world.

Random fractal forgeries: from mountains to music

Figure 1, like all of my colour illustrations, is an example of a fractal forgery. It is actually composed of three separate forgeries: the fractal landscape in the foreground, its fractally distributed craters, and the generalization of Brownian motion onto a sphere rising in the background. The objects represented in the images exist only as arrays of random numbers in a computer. Yet, the resemblance to the real world is unmistakable. My success with these forgeries and the extent to which they capture some of the essential characteristics of nature is due to a new branch of mathematics, *fractal geometry*,[1] as conceived and developed by Benoit Mandelbrot.

According to Galileo (1623),

> "Philosophy is written in this grand book—I mean Universe—which stands continuously open to our gaze, but it cannot be understood unless one first learns to comprehend the language in which it is written. It is written in the language of mathematics, and its characters are triangles, circles and other geometrical figures, without which it is humanly impossible to understand a single word of it; without these, one is wandering about in a dark labyrinth."

Galileo was, of course, philosophically correct: in order to understand or simulate nature, one must first comprehend its language. To scientists, mathematics is the language of nature and geometry is the appropriate dialect for understanding and manipulating shapes. Galileo was, however, wrong in terms of the dialect needed. Unfortunately, the inability of Euclidean geometry (the specific dialect to which he referred) to accurately describe the natural world has been overlooked until recently. In retrospect it is obvious that, "clouds are not spheres, mountains are not cones, coastlines are not circles, and bark is not smooth, nor does lightning travel in straight lines".[1] Fortunately, there is now a geometry that is appropriate for the irregular shapes of the real world. It is the goal of this paper to provide a brief introduction to and acquaintance with fractal geometry through the accompanying forgeries.

Figure 1 Fractal planetrise. A variation of Brownian motion on a sphere (D=2.5) above a fractally cratered random Gaussian fractal surface (D=2.2).

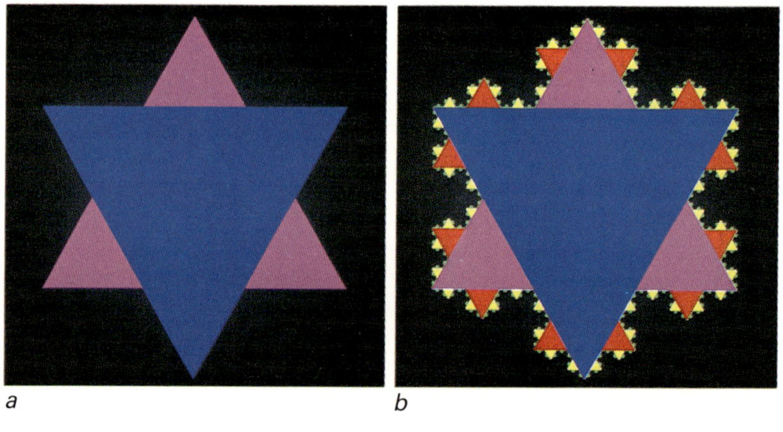

Figure 2 The exactly self-similar von Koch snowflake curve. *a*, The first stage in the construction where each outer side is divided into three equal parts and an equilateral triangle is added to the middle third. *b*, Later stages in the construction where the outer perimeter has a fractal dimension $D = \log(4)/\log(3)$.

Mathematical monsters

The turn of the century coincided roughly with an upheaval in the world of mathematicians. Their minds conceived of strange monsters seemingly without counterpart in nature: Cantor sets, Weierstrauss functions, Peano curves, curves without derivatives and lines that could fill space. Having once discovered these monsters (and congratulated themselves on a creativity superior to nature) they banished the beasts to a mathematical zoo. They could imagine no use for, or interest in, their creations by natural scientists.

Nature, however, was not so easily outdone. As shown by Benoit Mandelbrot, a mathematician at IBM's Thomas J. Watson Research Laboratory, not only do many of these "monsters" have counterparts in the real world, but they form some of the basic structures in the language of nature's irregular shapes: the fractal geometry of nature.

Fractals (a word coined by Mandelbrot in 1975) have blossomed tremendously in the past few years and have helped reconnect pure mathematics research both with the natural sciences and with computing. Within the past 5–10 years fractal geometry and its concepts have become central issues in most of the natural sciences: physics, chemistry, biology, geology, meteorology, and materials science. At the same time, fractals are of interest to graphic designers and filmmakers for their ability to create new and exciting shapes and artificial but realistic worlds. Computer graphics has played an important role in the development and rapid acceptance of fractal geometry as a valid new discipline. The computer rendering of fractal shapes leaves no doubt of their relevance to nature.

What then are fractals? How are they different from the usual Euclidean shapes? The answer begins with one of the early mathematical monsters: the von Koch snowflake curve (first proposed in 1904).

Self similarity

Figure 2 illustrates the iterative or recursive procedure for constructing a fractal curve. Beginning with an equilateral triangle, each of the outer sides is divided in three equal parts. Each middle portion serves as the base of a new smaller added triangle as shown in Figure 2*a*. With the next stage each of the new sides is again divided in three, with the centre serving as the base of new, even smaller, triangles. The procedure, repeated over and over yields the beautiful snowflake shown in Figure 2*b*. The outer boundary has, however, some highly unusual properties. At each stage in its construction the length of the boundary increases by a factor of 4/3. Thus, the limiting boundary crams an infinite length into a finite area of the plane without intersecting itself. The boundary has detail on all length scales. Indeed, the closer one looks, the more powerful the microscope one uses, the more detail one finds. More important, the boundary possesses an exact *self-similarity*. Each small portion of the boundary, when magnified, can reproduce exactly a larger portion. The boundary is said to be invariant under changes of scale.

This property of self-similarity or *scaling* is one of the central concepts of fractal geometry. It is closely connected with our intuitive notion of *dimension*. An object we consider to be one-dimensional, a line segment, for example, also possesses a similar scaling property. It can be divided into N identical parts each of which is scaled down by the ratio $r=1/N$ from the whole. Similarly, a two-dimensional object, such as a square in the plane, can be divided into N self-similar parts each of which is scaled down by a factor $r=1/N^{1/2}$. A three-dimensional object like a solid cube may be divided into N little cubes each of which is scaled down by a ratio $r=1/N^{1/3}$. With self-similarity the generalization of dimension is straightforward. A D-dimensional self-similar object can be divided into N smaller copies of itself each of which is scaled down by a factor r where

$$N = 1/r^D$$

Conversely, given a self-similar object of N parts scaled by a ratio r from the whole, its *fractal dimension* is given by

$$D = \log(N)/\log(1/r).$$

The fractal dimension, unlike the more familiar notion of Euclidean dimension, does not have to be an integer. Each side of the von Koch snowflake of Figure 2 is composed of four segments each of which is scaled down by a factor of 1/3 from the original side. Its fractal dimension $D = \log(4)/\log(3)$ or about 1.26. This non-integer dimension, greater than one but less than two, reflects the unusual properties of the snowflake boundary. It somehow fills more of space than a simple line ($D=1$), but less than a section of the plane ($D=2$).

Although objects in nature rarely exhibit exact self-similarity, they do often possess a related property, *statistical self-similarity*. Perhaps the simplest example is a coastline. As with the von Koch snowflake, the closer one looks at a coastline, the more detail one sees. If one were trying to measure the length of the coastline, the more carefully one follows the smaller wiggles, the longer it becomes. Moreover, each small section of a coastline looks like (but not exactly like) a larger portion. This concept of statistical self-similarity can also be quantified in terms of fractal dimension as above. In measuring the length of a coastline with a ruler of size r, the total length equals the ruler size r times the number of steps of size r, $N(r)$, taken in tracing the coast

$$\text{Length} = r \times N(r).$$

As with the snowflake, $N(r)$ varies as $1/r^D$ and

$$\text{Length varies as } r \times 1/r^D = (1/r)^{D-1}.$$

The variation of apparent coastline length with ruler size has been studied by Richardson. Real coastlines can, in fact, be characterized by fractal dimensions D of about 1.2 to 1.25, very close to the log(4)/log(3) of the Koch snowflake coastline.

The property that objects can look similar while at the same time different in detail at different length scales, is the central feature of all of the random fractal forgeries presented here. Figure 3 shows a series of views, at increasing magnification, of a computer generated fractal landscape with just such a statistically self-similar coastline. The upper left image shows a section of the landscape. A small portion, as indicated by the white box in the second frame, is then magnified to fill the third frame. In each succeeding frame, a small portion about the coastline is magnified to fill the next frame. This magnification process can be continued indefinitely. Here the magnification is carried to a total of more than 16 million and the final frame repeats the starting view. Each frame is clearly a magnification of a portion of the previous, yet each frame also looks like a different part of the same landscape *at the same magnification*. This is the statistical self-similarity of a random fractal. It is random in the sense that (unlike the snowflake) a large scale view is insufficient to predict the exact details of a magnified view. The way in which the detail varies as one changes length scale is once again characterized by a fractal dimension. The irregular fractal surface is more than a simple surface (D=2) and the sample shown in Figure 3 has D=2.2. The fractal dimension of the coastline is one less than that of the surface itself. Here the coastline D=1.2.

Fractal landscapes

The fractal landscapes were produced by adding random irregularities to a surface at smaller and smaller scales similar to the process of adding smaller and smaller triangles to the snowflake. As shown in Figure 4, one begins with a relatively smooth surface and then adds smaller and smaller hills until the effects of the individual hills and the underlying regular grid become invisible. Most fractal forgeries were produced the same way: effectively by adding more and more detail as one goes to smaller and smaller scales. In many cases special computational techniques, such as the Fast Fourier Transform reduced the large amount of computation required.

The fractal dimension determines how the relative amounts of detail or irregularities at different distance scales is varied. Surfaces with a larger D seem rougher. This effect is illustrated in Figure 5 which shows the 'same' surface but with different fractal dimensions. Figure 5*a* has a relatively low

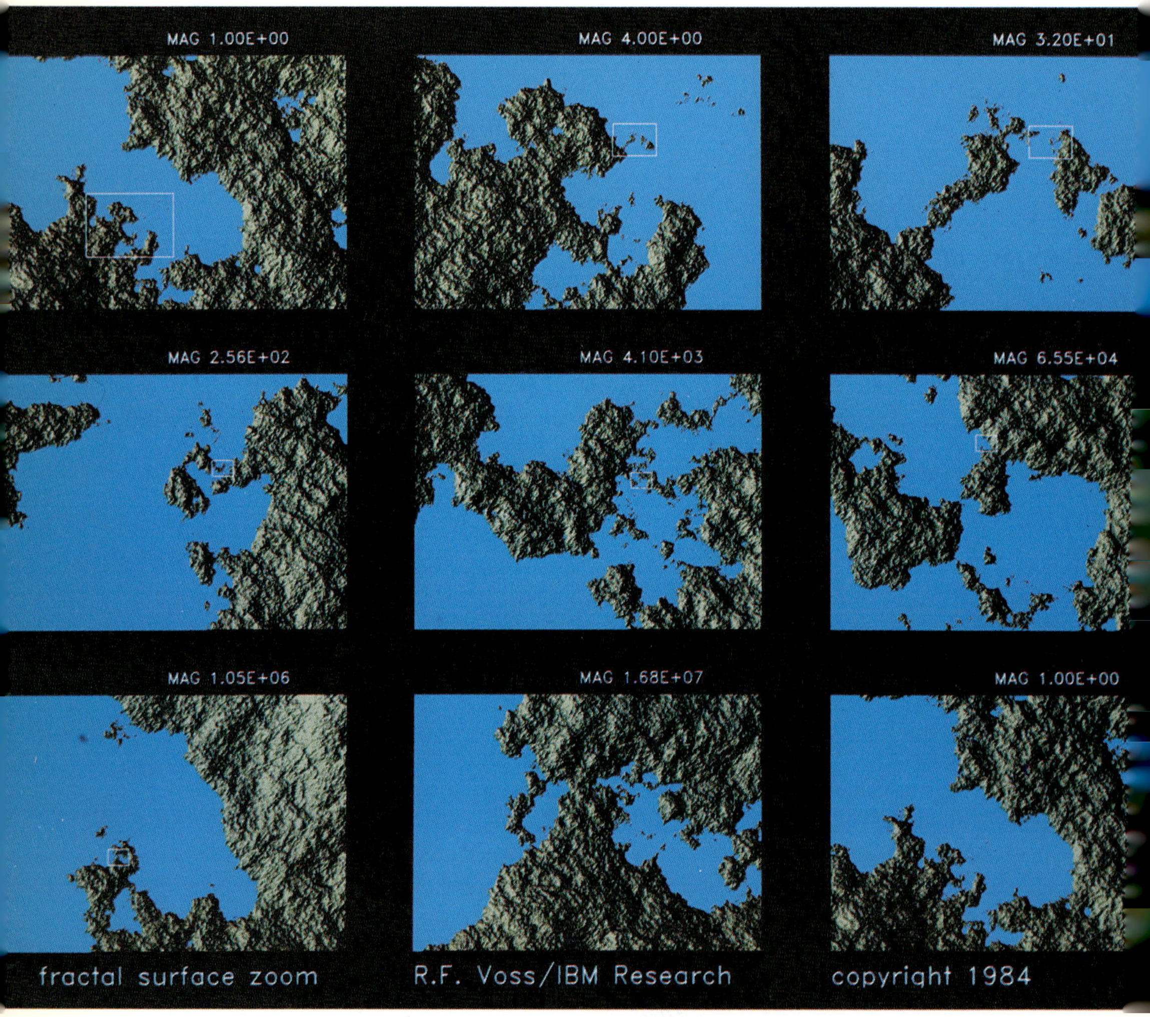

Figure 3 Zoom sequence of a statistically self-similar fractal landscape (D=2.25). Each succeeding picture shows an enlargement of the framed portion of the previous image. As the surface is magnified, a small portion looks similar to (but not exactly the same as) a larger portion. The total magnification corresponds to 16 million.

fractal dimension for the surface, D=2.15, that is appropriate for much of the Earth. An increase in the fractal dimension to D=2.5 is shown in Figure 5*b*. A further increase to D=2.8 gives an unrealistically rough surface for an actual landscape. Fractal geometry specifies only the height variations of the landscape. The colouring was primarily based on height above water level and local slope.

The image in Figure 5*a* does not resemble many of the valleys on the Earth's surface which are significantly eroded. A flat bottomed basin can, however, be approximated mathematically by simply taking the height variations and scaling them by a power-law. This process is illustrated in Figure 6 where the original landscape of Figure 6*a* is "cubed" to produce Figure 6*b*. The effect of such a power greater than 1 is to flatten the lower elevations near the water emphasizing the peaks. Scaling with a power less than one has the opposite effect of flattening the peaks while increasing the

Figure 4 The creation of a fractal landscape by adding detail at successively smaller (here by a factor of 2) sizes. The fractal dimension D specifies the relative amount of large to small scale irregularities.

Figure 5 The differing visual effect of changing the fractal dimension *D* for the same landscape. *a*, D=2.2; *b*, D=2.5; *c*, D=2.8. As *D* increases, the perceived surface roughness also increases.

Figure 6 Fractal landscaping by scaling the surface height variations (relative to water level) by a power law. *a*, A 'virgin' Gaussian random fractal landscape. *b*, Heights to the power 3. *c*, Heights to the power 1/3.

steepness near the water. A cube-root processing of Figure 6*a* is shown in Figure 6*c*. This forgery gives the impression of river erosion into an otherwise fairly smooth plain.

In order to give the impression of a lunar landscape, as shown in the foreground of Figure 1, it is necessary to add craters to the 'virgin' fractal landscape of Figure 5*a*. Each crater is circular with a height profile similar to the effect of dropping marbles in mud. The trick in achieving realism is to use the proper distribution of crater sizes. For the Moon, the actual distribution is fractal or power-law, with many more small craters than large ones. The computer simulation is shown, viewed from above, in Figure 7. This is the same surface that forms the foreground of Figure 1.

The creation of the rising fractal planet in Figure 1 follows a different process that is closer to an actual model for the evolution of the Earth's surface. The surface we are approximating is a generalization of Brownian motion or a random walk on a sphere. In the same way that a random walk is the sum of many independent steps, the sphere's surface becomes the result of many independent surface displacements or faults. Each fault encircles the sphere in a random direction and divides it into two hemispheres which are displaced relative to each other in height. Inside the computer, the surface of the sphere is mapped onto a rectangular array similar to a flat projection map of the Earth. Figure 8*a* shows the surface variations, as represented by regions of different colour, both for the full and the flattened sphere after only 10 random faults. After 400 faults

Figure 7 A fractal landscape (the same as Figure 5*a*) with craters whose area distribution follow a fractal or power-law dependence with many more small craters than large ones.

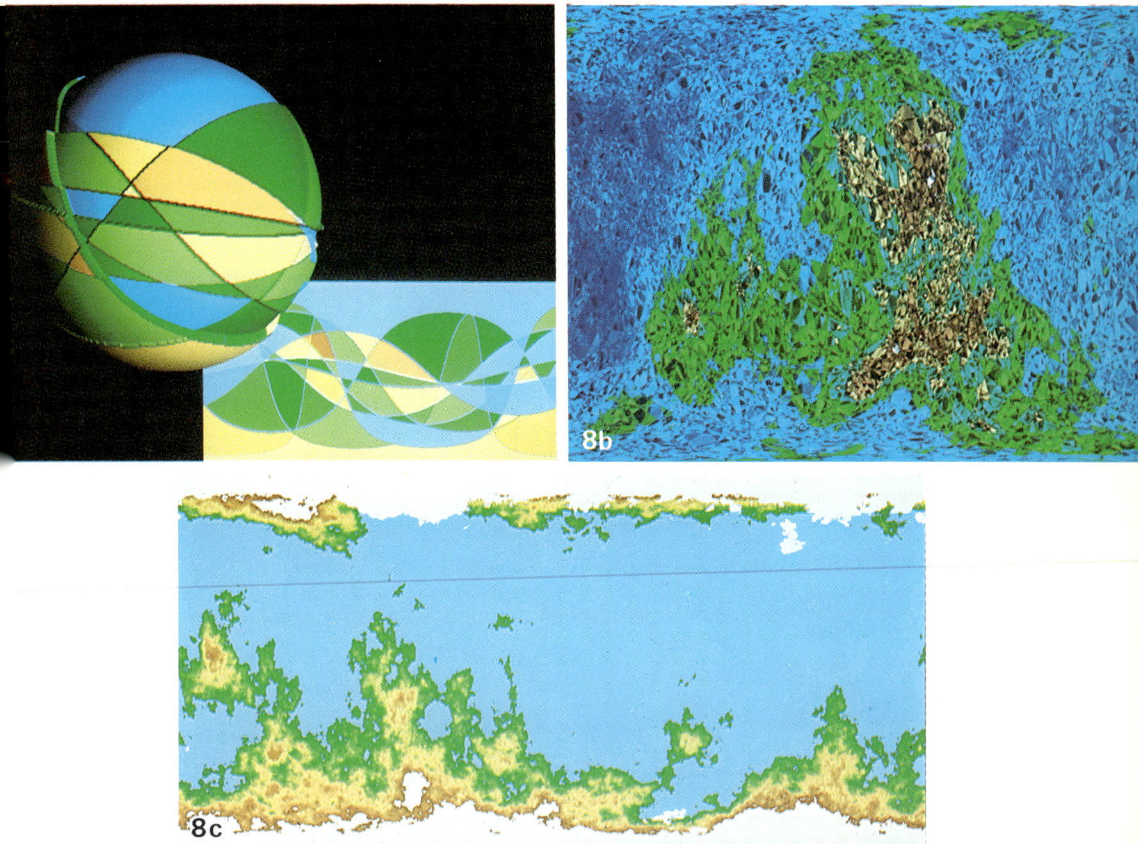

Figure 8 The stages of creating a fractal planet by adding random faults encircling the sphere. *a*, The surface height variations (as indicated by colour) after 10 faults shown both on the sphere and on a flat projection map. *b*, The surface map after 400 faults. *c*, The surface after more than 10,000 faults with added polar caps.

(Figure 8*b*) the sphere has developed a localized land mass and the effects of an individual fault are becoming lost. After more than 10,000 faults and the addition of polar caps, the sphere's surface, (Figure 8*c*) becomes a plausible planet. The fractal dimension of the resulting surface is D=2.5. Mapping the data of Figure 8*c* onto the sphere gives the fractal planet in Figure 1 or the enlarged view of the rotated sphere shown in Figure 9*a*. Remarkably, the same sphere is also a good forgery of a much more arid planet. Figure 9*b* shows the same data but with different colouring and without water.

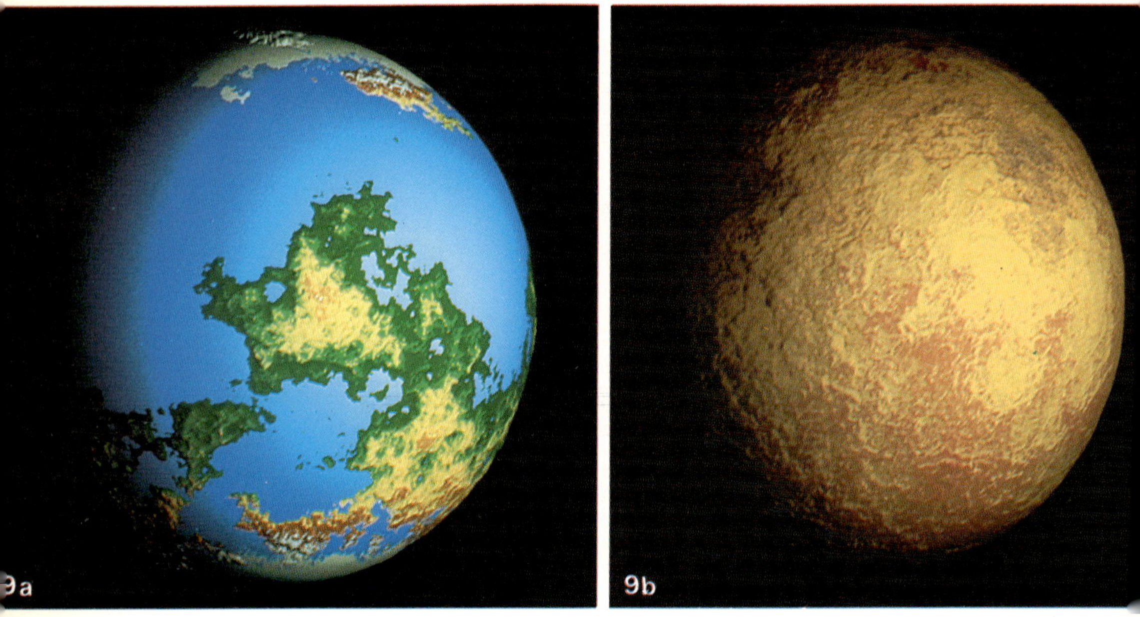

Figure 9 The Brownian fractal planet mapped back onto the sphere (surface D=2.5). *a*, With an Earth-like water level and polar caps. *b*, Same planet without water.

Figure 10 Random fractals extended to density variations in three dimensions that scatter light similar to actual clouds (D=3.5). This forged cloud is placed above a forged fractal landscape.

Fractal clouds

All of the fractal forgeries shown up to now have been pretentious surfaces. The scaling irregularities raise their dimension from a topological value of 2 to a fractal value $2<D<3$. It is, however, also possible to generate fractals with a topological dimension 3 whose scaling irregularities raise the fractal dimension to $3<D<4$. Rather than landscapes, which consist of random heights as a function of two-dimensional surface position, such objects might represent a random temperature or water vapor density as a function of three-dimensional position. A good example is the fractal cloud shown in Figure 10. In this case, both the computation of the fractal data as well as the actual rendering of the image (allowing for light and shadow to partially penetrate the cloud) required much more computation than the previous landscapes.

Until now, I have focused on variations or randomness in space. The computer generated forgeries show the extent to which simple fractal constructions capture the essence of natural shapes. Changes in time, however, have many of the same qualities as changes in space.

Noise and music

To the physicist, unpredictable changes in time are known as *noise*. Graphical samples of typical noises are shown in Figure 11. To the left of each sample is a representation of its *spectral density*. The spectral density gives an estimate of the time correlations, or lack thereof, in the random process. All of these noises are also fractal and there is a direct relationship between the fractal dimension and the slope of the spectral density. Figure 11*a* shows the most random, a *white noise*. It could be produced by a random number generator and is completely uncorrelated from point to point. Its spectral density is a flat line, independent of frequency. Alternately, a white noise contains equal amounts of all frequencies (like a white light). Figure 11*c* shows a Brownian motion or a random walk, the most correlated of the three noise samples. It consists of many more slow (low frequency) than fast (high frequency) fluctuations and its spectral density is quite steep. It varies as $1/f^2$. In the middle (Figure 11*b*), is an intermediate type of noise known as $1/f$ noise because of the functional form of its spectral density. Although both white and $1/f^2$ noise are well understood in terms of mathematics and physics, $1/f$ noise, whose origin remains a mystery to scientists after more than 60 years of investigation, represents the most common type of noise found in nature.

There are no simple mathematical models that produce $1/f$ noise other than the tautalogical assumption of different time constants. Little is also known about the physical origins of $1/f$, but it is found in most complicated physical systems. $1/f$ noise is found in almost all electronic components from simple carbon resistors to vacuum tubes and all semiconducting

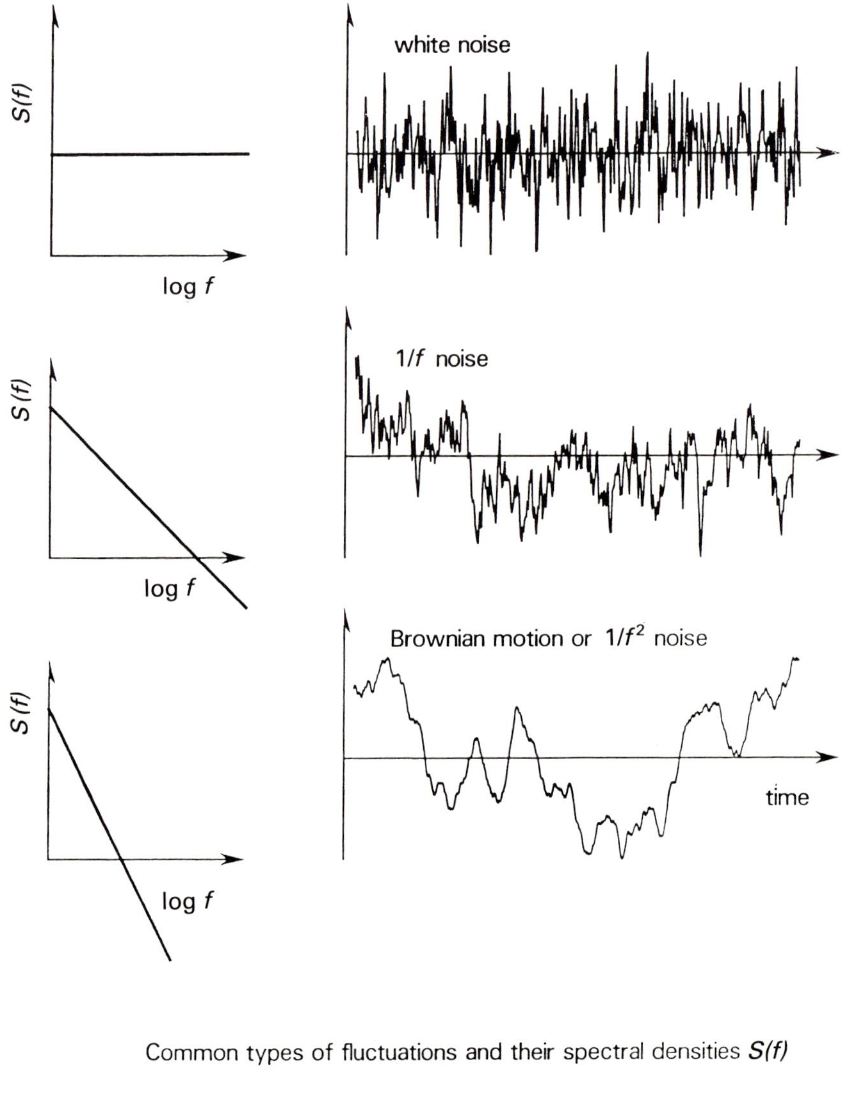

Figure 11 Samples of typical 'noises', the random variations of a quantity in time. *a*, White noise, the most random. *b*, 1/*f* noise, an intermediate but very commonly found type of fluctuation in nature, its origin is, as yet, a mystery. *c*, Brownian motion or a random walk. To the left of each sample is a graphical representation of that noises spectral density, a measurement technique for characterizing the time correlations in the noise.

devices. It occurs in all time standards from the most accurate atomic clocks and quartz oscillators to the ancient hourglass. It is found in ocean flows and the changes in yearly flood levels of the river Nile as recorded by the ancient Egyptians. It is found in the small voltages measurable across nerve membranes due to sodium and potassium flow. Even the flow of automobiles on an expressway gives $1/f$ noise.

One of my most exciting discoveries was the measurement that almost all musical melodies also mimic $1/f$ noise. Music has the same blend of randomness and predictability that is found in $1/f$ noise. If one takes a music score and draws lines between successive notes of the melody one finds a graph remarkably similar in quality to Figure 11*b*. Some of the actual experimental data are shown in Figure 2. There is little to distinguish these measured spectral densities for different types of western music from each other or from the $1/f$ noise shown in Figure 11*b*. This type of analysis seems insensitive to the different types of music. It emphasizes, instead, the common element in music and suggests an answer to a question that has long troubled philosophers. In the words of Plato, "For when there are no words (accompanying music), it is very difficult to recognize the meaning of the harmony and rhythm, or to see that any worthy object is imitated by them". Philosophers generally agreed on the imitative nature of the arts. It seemed obvious that painting, sculpture or drama imitated nature. But what does music imitate? The measurements suggest that music is imitating the characteristic way that our world changes in time. Both music and $1/f$ noise are intermediate between randomness and predictability. Like fractal shapes there is something interesting on all (in this case, time) scales. Even the smallest phrase reflects the whole.

It is, of course, possible to use fractals, in this case as $1/f$ noise, to forge music as well as landscapes. Figure 13 shows samples of 'music' generated from the three characteristic types of 'noise' shown in Figure 11. Although none of the samples in Figure 13 correspond to a sophisticated composition of a specific type of music, Figure 13*b* generated from $1/f$ noise is the closest to real music. Samples of this $1/f$ music sound recognizably musical, but from a foreign or unknown culture.

Progress

Man has always been confronted with a world filled with seemingly complex irregular shapes and random fluctuations. In the quest for understanding, natural science has progressed by concentrating only on the simplest of systems. In the process, it has moved away from the direct experience of nature to the electronically instrumented laboratory. After all, who could describe, let alone understand, the profile of a mountain or the shape of a cloud? With fractal geometry, the quest can return to the everyday natural world.

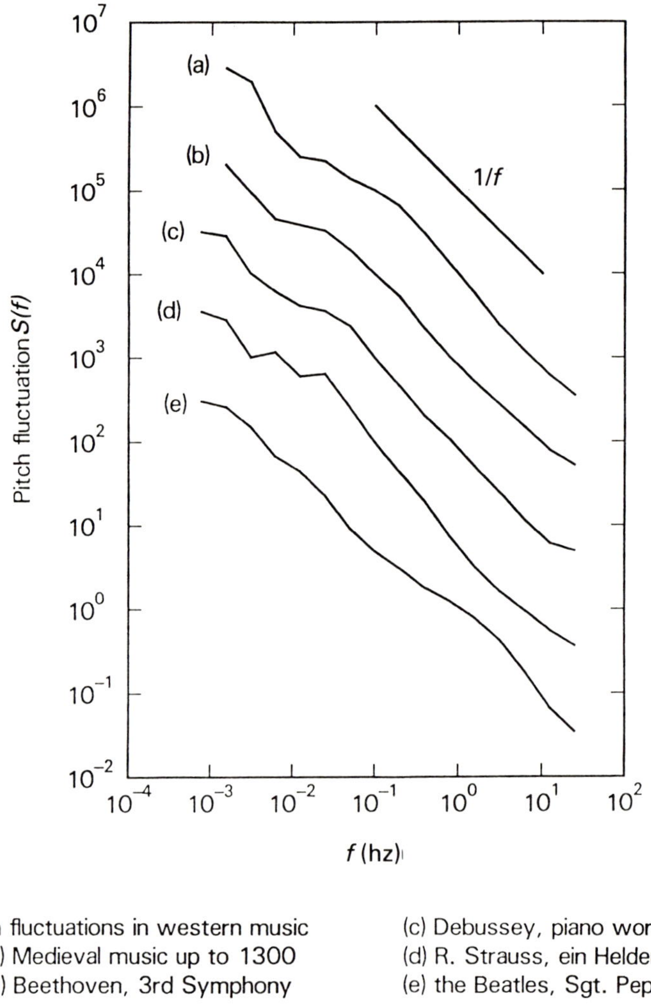

Figure 12 Noise measurements of the pitch variations in various types of music showing their common correlations as $1/f$ noise.

References

1. Mandelbrot, B., *The Fractal Geometry of Nature*, Freeman, New York (1982).
2. Gardner, M., 'White and brown music, fractal curves, and one-over-f noise', *Scient. Am.*, 16 (1978).
3. Voss, R. F., and Clarke, J., '"1/f noise" in music and speech', *Nature, 258*, 317–318 (1975).

(a)

(b)

(c)

Figure 13 Sample of stochastically-composed music based on the different types of noises shown in Figure 11*a*, 'White' music is too random. *b*, $1/f$ music is the closest to actual music (and most pleasing). *c*, 'Brown' or $1/f^2$ music is too correlated.

Professor Bob Churchhouse CBE

Bob Churchhouse was appointed the first holder of the Chair in Computing Mathematics at University College, Cardiff, in 1971. After graduating from Manchester University with a First Class Honours degree in Mathematics, he went to Trinity Hall, Cambridge and carried out research in Number Theory under the supervision of Professor L.J. Mordell, obtaining his PhD in 1952. From 1952 to 1963 he was employed at GCHQ, when he left to join the newly-founded Atlas Laboratory as Head of the Programming Group.

He has served on the Councils of the London Mathematical Society and the Institute of Mathematics and its Applications, being Vice President of the latter from 1981 to 1983. He was, in 1965, a member of the Flowers Committee, which set the pattern for computing in the Universities, and from 1975 to 1982 a member of the Computer Board for Universities and Research Councils, being Chairman from 1979 to 1982. His Fellowships include the British Computer Society, the Institute of Mathematics and its Applications, and the Royal Astronomical Society.

His main research interests are concerned with the application of computers in mathematics, information retrieval, and literary and linguistic studies.

Uncertainty in mathematics and computing

This discussion will not be concerned, except in a very tenuous way, with probability theory; but I have taken this opportunity, when we have a large number of people in a room, to conduct an experiment related to a well-known result of probability theory. The 'well-known result' is that if there are 23 randomly chosen people in a room there is an evens chance that there is a pair among them who have the same birthday. Many mathematicians are surprised when they first hear this and most other people find it barely credible. The explanation is quite simple: from 23 people we can form $\frac{1}{2} \times 23 \times 22 = 253$ pairs; the probability that any particular pair have the same birthday is $1/365$ so the probability that none of the 253 pairs have the same birthday is

$$\left(1 - \frac{1}{365}\right)^{253} \doteqdot e^{-\frac{253}{365}} \doteqdot 0.5$$

so that the probability that at least one pair have the same birthday is $(1 - 0.5) = 0.5$.

Now there are 240 people attending this lecture, so what would be expected in the way of common birthdays in such a case? The calculation is not difficult; the prediction is that we would expect (to the nearest integer): 1 group of 4 with the same birthday, 9 groups of 3, 41 groups of 2, leaving 127 with unique birthdays.

Of the people attending this conference, 1 group of 4 have the same birthday, 9 groups of 3 have the same birthday, 48 groups of 2 have the same birthday and 113 have unique birthdays.

I will now turn to the topic of my talk.

Axioms and theorems

Uncertainty is not a word that one normally associates with most branches of mathematics. Branches of mathematics are based on sets of axioms from which deductions are made; a correct set of deductions leads to a theorem; thus from the axioms of arithmetic it not only follows fairly quickly that $2 + 2 = 4$ but also after a considerably longer chain of deductions that every positive integer is the sum of four squares. Of course, it would be tedious if we had to write down the proof of every theorem starting from the axioms and so when a theorem has been proved it becomes, in effect, an extension of the axioms and may be used to provide a short-cut to the proof of further theorems.

Thus the corpus of the subject steadily expands with theorems deduced from combinations of earlier theorems, which themselves follow from combinations of earlier theorems and so on, until eventually everything can be traced back to the axioms. The entire structure rests on the axioms and is completely sound provided that every theorem in the structure is really true. If a single erroneous result is included, those parts of the structure which depend on it are in danger of collapsing; for if we believe that a certain false result is true we have effectively saddled ourselves with a binary logic in which 'zero' and 'one' are the same and in such a system *everything* is deducible, that is everything is both true *and* false. (A well-known mathematician once said that if we assumed $0 = 1$ we could prove anything, so someone asked him to prove that he was the Pope. "Since $0 = 1$" he said "$0 + 1 = 1 + 1$, that is $1 = 2$; so if the Pope and I are 2 people we are therefore 1 person, i.e. we are the same.")

Now anyone who treats 0 and 1 as the same deserves whatever consequences follow, but it is possible to get perilously close to that situation without realizing it. Nearly 30 years ago I programmed a computer, in machine code, of course, and discovered that this computer had two forms of 'zero', which it recognized separately as 'plus zero' and 'minus zero'. 'Plus zero', for example, was the result of subtracting 5 from 5 whereas 'minus zero' was the result of adding 5 to minus 5. If we asked this computer whether 'plus zero' and 'minus zero' were the same it replied that they were not. In the logic of this machine, therefore, there were two, unequal, forms of zero. From a logical point of view it is not immediately obvious whether having two unequal zeros is as disastrous as assuming that zero equals one, but one would expect that it would lead to some bizarre arithmetic at least and might well lead to a total logical and arithmetical collapse. The consequences of such an eccentric arithmetic for banks (unable to balance their books) and airlines (number of passengers getting off aeroplanes not equal to number getting on) are easily imagined.

Uncertainty in mathematics and physics

To return to the history of pure mathematics: until the 1930s it was commonly believed that any true result must be eventually deducible from the axioms, although the chain of deductions might be so long that no-one could complete them in a lifetime, but in 1931 Kurt Gödel proved that in any system there will always be statements which can neither be proved nor disproved.

In the case of ordinary arithmetic, Gödel proved even more *viz.* that there are statements about the integers which are, in fact, true but which we can never prove. Any theologian might like to consider whether in the case of a statement which *is* true but undecidable from the axioms, would God know that the statement is true and would he be able to prove it, and, if not from the axioms, how?

These results astonished not only the mathematical world but also lots of other people, including philosophers who had become very excited a few years earlier by Heisenberg's Uncertainty Principle in Physics. In its simplest form, Heisenberg's Uncertainty Principle states that we cannot measure *both* the position and momentum of an electron simultaneously with arbitrary accuracy; or, more formally, the product of the errors in position and momentum is essentially constant, so as we improve the accuracy in one aspect we necessarily lose accuracy in the other. Heisenberg put forward his Uncertainty Principle in 1926, five years before Gödel produced his great theorem, which might be regarded as an Uncertainty Principle of Pure Mathematics; not a quantatitive one admittedly, but with uncertainty as its essence. So far as I know, this close juxtaposition in time of these two landmarks was a pure coincidence, Heisenberg clearly could not have known of Gödel's result, and there is no evidence to suggest that Gödel was in any way influenced by Heisenberg's.

Why did the philosophers get so excited about Heisenberg's Uncertainty Principle? Presumably because hitherto they had regarded science in general and physics, in particular, as 'totally deterministic' and the introduction of 'uncertainty' into such a discipline was sensational. How much more sensational then when 'Uncertainty' appeared in pure mathematics, the totally pure, incorrupt, logical Queen of all the Sciences? Whether most philosophers understood either the Heisenberg Uncertainty Principle or Gödel's Theorem is itself very uncertain.

To return briefly to Gödel, however: it would, of course, be even more interesting if we could say whether any given statement is decidable or not, but because we cannot do this long-standing conjectures such as Fermat's Last Theorem and the Riemann Hypothesis, which have been the subjects of thousands of papers during the past hundred years or so, and which are still unproved after more than 350 and 120 years respectively, will continue to engage the attentions of mathematicians.

The empirical evidence for the truth of both these conjectures and, indeed, for several more, such as Goldbach's Conjecture, that every even number is the sum of two primes, or the conjecture that there exist an infinite number of pairs of primes, is very strong, and most mathematicians believe them to be true, but we have to face the possibility that any of them may be true but unprovable.

Uncertainty in computing

There is an analogous theorem relating to computer science. It would be invaluable to computer manufacturers, software houses and, indeed, computer programmers, if an algorithm could be constructed which would generate a comprehensive set of test input data for *any* given programme, i.e. a set of data such that if the program produces the correct output for all the input data in the set then it must produce the correct output for any input whatsoever. Unfortunately, the so-called 'Recursion Theorem' shows that no such general algorithm can exist. This does not mean, of course, that we cannot prove that any *specific* program is correct, merely that no proof of correctness based on test cases can be constructed by a program which takes arbitrary programs as its input.

Now although the Heisenberg Uncertainty Principle of Physics, the Gödel Incompleteness Theorem of Mathematics, and the Recursion Theorem of Computer Science are fundamentally quite different they do have one property in common: they all tell us that there is a limit to what we can do, no matter how clever we are, or how hard or for how long we try. Is it possible then that there are other examples of this type of theorem: i.e. theorems that tell us that there is a limit to what we can do in certain disciplines? This might well be so and in support two examples can be given from computation. The first can be proved mathematically; the second is empirical. Both take a form reminiscent of the Heisenberg Principle.

When we carry out any computation on a computer there is, of course, a limit to the precision that we can obtain, determined by the word length of the machine. This is the trivial, or obvious limitation. For this example, incidentally, the possibility that multi-length arithmetic facilities might be available is irrelevant, the limitation that I am about to describe will appear sooner or later. Suppose that on a machine having, say, 24 binary-place precision we wish to carry out a simple numerical integration, that is estimate the area under a curve. You might think that by using one of the well-known formulae (such as the Midpoint Rule) with a sufficiently small interval size, we could estimate the area to within an error of 2^{-24}. It is salutary to try doing this on a computer. As the interval size is reduced, the error at first falls exactly as it should but then it begins to rise and the harder we try the bigger the error becomes. In consequence, we might well find, as I did, that the maximum precision achieved is not 24 binary places but about 19. The explanation is that the error we get is a combination of two different kinds of error: the *truncation error*, which gets smaller as we try harder, and the *rounding error*, which gets bigger. The magnitudes of these two errors are related by an approximate formula of the type

$$(\text{Truncation error}) \times (\text{Some power of the rounding error})$$

is approximately constant. This form is reminiscent of that of the Heisenberg Principle.

The value of the constant and the power of the rounding error in this relationship depend upon the function, the method and the machine, but it can be predicted. It would be possible, at least in some cases, to build the calculation of this value into the computer program so that the optimal choice of interval length to minimize the overall error would be made.

The second example illustrates a facet of computer-based information retrieval systems. Suppose we have a large database of titles of articles in various scientific journals and that we wish to obtain details of all articles on a certain topic. To do this we put in a request of the form:

"Find all articles which satisfy condition C_1 *or* condition C_2 *or* condition C_3 *or* . . . *or* condition C_n." Such a request is very 'loose' and if n is even moderately large, many articles will be retrieved. Most of them will be of no interest, but most of the interesting articles will be retrieved. In a typical case, suppose there are 100,000 articles in the database of which 32 really are of interest. A request of this 'loose' type might retrieve about 200 articles, including, say, 24 of those that are really interesting. In such a case we say that the *Recall ratio* is 24/32 and that the *Precision ratio* is 24/200. The product of these ratios is 0.09. Alternatively we might put in a 'tight' request; that is, one of the type:

"Find all articles which satisfy condition C_1 *and* condition C_2 *and* . . . *and* condition C_n." In this case if n is moderately large very few articles will be retrieved, but most of them will be of interest. For example, we might retrieve only 50 articles in this case of which, however, 12 may be of interest. The Recall ratio will then be 12/32 and the Precision ratio will be 12/50; the product of these ratios, however, is again 0.09.

Now the figures just given are purely hypothetical but, in my experience, in any particular information retrieval system there is an empirical 'law' of the type: (Recall ratio) $\times$ (Precision ratio) $\doteqdot$ Constant.

The "Constant" depends upon the database, the keyword and the logical system, but when we try to retrieve as above it seems that Recall and Precision are negatively correlated in a logarithmic sense. Of course, in practice a well-designed system will respond to a 'loose' question with a statement of the type: "If you insist on phrasing your question like that you will get 6,462 articles"—hence inviting the user to 'tighten' the question; but the basic characteristic property—a Heisenberg type Uncertainty—is still there.

In conclusion then it seems that we have uncertainty in more disciplines than is generally realized and that it takes more than one form but in at least some cases it takes a form which might be stated as: "As we improve the accuracy of the measurement of one quantity so we inevitably decrease the accuracy of another."

Professor David G. Kendall

David Kendall is Professor of Mathematical Statistics, and Fellow of Churchill College, University of Cambridge.

After his early studies under E.A. Milne, his career included Lecturer at Magdalen College, Visiting Lecturer at Princeton, USA, DSc (Oxford) and Dr. honoris causa (Paris). His is a Fellow, and Member of Council, of the Royal Society, and was awarded the Royal Society Sylvester Medal in 1976. He has held the posts of President of the London Mathematical Society (1972–74), the Bernoulli Society (1975), and the Mathematics and Physics Section of the British Association (1982).

His wide interests in the applications of mathematics and statistics has also been reflected and recognized in his work with the Yorkshire Archaeological Society (and one of his books "Mathematics in the Archaeological and Historical Sciences"), the Weldon Medal Prize and Medal for Biometry, and the Silver and gold Medals of the Royal Statistical Society.

Chance or design?*

E.A. Milne was born in Hull in the East Riding of Yorkshire in 1896, and died in Dublin in 1950 at the relatively early age of 54. I first met him within a few days of my arrival in Oxford in 1936. Through the kindness of Frank Twyman of Adam Hilger Ltd I had been given a letter of introduction to the Director of the University Observatory, H.H. Plaskett, who promptly invited me to the Observatory Colloquium (4.30 on Wednesdays, with a cup of tea, and cake also if fresh offprints by an Observatory author had arrived that week). This I attended without fail throughout my three years as an undergraduate. On that first day Milne immediately came over and talked to me, and continued to show me great kindness until his death, as indeed did Plaskett also. I need not say that it was somewhat alarming to find one's self thrust into such company, but Milne's warm personality and infectious excitement about the latest discoveries quickly melted away most of my shyness. After that I also attended his own seminar regularly, and was a frequent visitor at 19 Northmoor Road. I remember helping to decorate the house and garden with bunting for the Coronation in 1937, with the assistance of two delightful small girls to whom I send warm greetings today.

Milne was a vivid lecturer and always communicated at least his enthusiasm to the audience. His undergraduate course on Vectorial Mechanics was a *tour de force*, but of his seminar lectures I remember now only his insistence that 'mass is just a constant of integration'. This seminar went on from year to year without recapitulation, having started I suppose in 1929 or thereabouts. If anything notable had happened in the preceding week (like the death of Rutherford, or Dirac's letter to *Nature* on 'big numbers') Milne would abandon the course and give his immediate reactions. I recall his saying that there were good days and bad days with Rutherford. The good days were immediately recognizable by his coming into the Lab whistling "Onward, Christian Soldiers."

---

*The Milne Lecture, University of Oxford, 1983.

My own research interest was then in such matters as radiative equilibrium and the Milne integral equations (of the Wiener–Hopf type subsequently to prove of importance in statistics), and I was encouraged in this by Milne, Plaskett and E.C. Titchmarsh. Milne was very proud of his training as an analyst by G.H. Hardy. It was Hardy who in 1916 introduced Milne to A.V. Hill, and so became responsible for Milne's membership of the famous Anti-Aircraft Research group organized and directed by Hill. Later (1939–1944) he became a Member of the Ordnance Board, an extraordinary body of mediaeval origin, and brought me also within its orbit in a very minor capacity. One piece of experimental work which I carried out rather against the wishes of some of my superiors ultimately found its way on to Milne's desk as an unsigned report, and was then used by him in the context of very remarkable but still unpublished work which involved some quite delicate analysis. Milne himself considered this to be the best thing he had ever done. When later Milne realized that it was my data he had used, he was delighted, and so was I.

Throughout the war Milne continued writing his constant stream of papers on kinematic relativity. When I asked him how he found the time, the answer was: 'I write them before breakfast'.

After the war I returned to Oxford as a novice Fellow of Magdalen, and Milne helped me to step out of the egg-shell into academic life. By then my interests had swung away from astronomy towards statistics, but I was very pleased to be able to tell Milne that Giordano Bruno too had espoused the Cosmological Principle. "If so," Milne later wrote, "I am in good company. Bruno was burnt at the stake."

In 1950, Milne was President of the Royal Astronomical Society, and it may have been by his wish that the Society met that year in Dublin; what should have been a very happy occasion was saddened by his collapse and death. These events impressed themselves on my mind, and when, many years later, I found myself President of the London Mathematical Society, I took the opportunity of encouraging it too to have an Irish meeting, which was much enjoyed and appreciated by all. Reflecting on this again, it occurs to me that it might be thought very appropriate for the Milne Society to meet in Dublin from time to time.

It is a privilege and a pleasure to be invited to give this lecture in memory of a great man who was also a good friend.

Statistics and geometry are obviously necessary tools for the study of the cosmos, but it has not been usual to apply them together. Classical statistics was primarily tailored to fit problems in Euclidean space, and despite two early pioneer papers by von Mises[1] and Fisher[2] it is only recently that statistical problems on manifolds have received systematic attention. When one passes from statistics on manifolds to stochastic motions on manifolds, then the global as well as the local geometry

Q. Jl R. astr. Soc. 25, 147–156 (1984).

Plate 1 E.A. Milne as I remember him. (This photograph was taken in May 1939 at Williams Bay, Wisconsin by Professor S. Chandraseskhar, FRS, with whose kind permission it is reproduced here.)

Plate 2 E.A. Milne as FRS. (From a photograph in the possession of Milne's family.)

Plate 3 E.A. Milne as Lieutenant RNVR. (From a photograph in the possession of Milne's family.)

becomes very important. Thus it will not suffice to employ analysis posing as geometry in a dress of upper and lower suffixes. The global geometrical structure becomes itself of prime significance.

My own interest in such problems has an archaeological origin. When one maps the positions of a given class of archaeological sites, there will always be some who claim that they are positioned on a framework of unseen straight lines (or perhaps circles, etc.). Here, of course, it is understood that each such straight line must, to avoid trivialities, pass through (or nearly through) at least three points, so the basic question becomes one of deciding how many triplet collinearities one can expect to find by chance in a given set of n points in the plane. To make sense of the question one must be precise about the 'nearly'. As a first shot we could say that three points are ε-collinear when they form a triangle with largest angle greater than $\pi-\varepsilon$. Here ε is to be a small angle, given in advance. But if we take this view then what we are doing is to consider the whole set of

$\binom{n}{3}$ triangles and ask whether an unduly large number of them have a *shape* which meets the test of ε-collinearity. From this point onwards one can proceed in various ways,[3] but an attractively general procedure is to carry out the investigation within a general theory of shape[4] for sets of k points in m dimensions, where in the example just mentioned k is 3 and m is 2.

Of course, it is essential to make clear what we mean by 'shape'. So let us say that shape is what is left when the effects of location, size, and rotation have been filtered out. Even this is unsatisfactory, however, because 'size' could be assessed in various ways. What seems to me natural is to assess size (or rather 'size squared') by calculating the sum of the squares of the distances of the points from their centroid. This has the advantage of being meaningful for all k and all m. A quite different theory, developed independently by my friend Ruben Ambartzumian (of an astronomically famous family), assesses size when $k = 3$ and $m = 2$ by the length of the perimeter of the triangle formed by the three points. It is not obvious how that should be generalized to larger values of k.

If we proceed as I have suggested here, we find that in two dimensions each shape (that is, each equivalence class of k-ads in $\mathbb{R}^2$ equivalent under displacements, rotations, and magnifications) can be identified as a point on a $2k-4$ dimensional manifold called $\mathbb{C}P^{k-2}$ (complex projective space). It can be shown that the representation is topologically correct, and indeed more is true. To understand what this means we need to introduce the further idea of the distance between two such shapes. Here the inspiration comes from classical mythology. We follow Procrustes (who so treated those unwise enough to accept his hospitality when travelling from the Peloponnese to Athens) by forcibly matching one k-ad against another by suitable rotations, translations, and changes of scale. (Procrustes went further, and allowed himself the additional operation of truncation, not appropriate here.) A coarse and natural measure of the discrepancy between two k-ads is the sum of the squares of the distances from each point of one k-ad to the correspondingly labelled point of the other. Obviously we ought to standardize that discrepancy for a common change of scale, so we divide it by the sum of the squares of the distances of all the $2k$ points from their pooled centroid. The ratio so formed is now dimensionless, and will have a minimum if we consider all allowable procrustean transformations of the two k-ads. If this minimum value is D, we define $\varrho = \arccos\,(1 - D)$, which is a distance-function, to be the distance between the two shapes. When this programme is carried out, our shape-manifold becomes a Riemannian one, and it turns out that as a Riemannian manifold it is actually isometric with the algebraic geometers' complex projective space up to the choice of a unit of length. Thus we have reduced the study of the shapes of k-ads in $\mathbb{R}^2$ to a classical geometric situation.

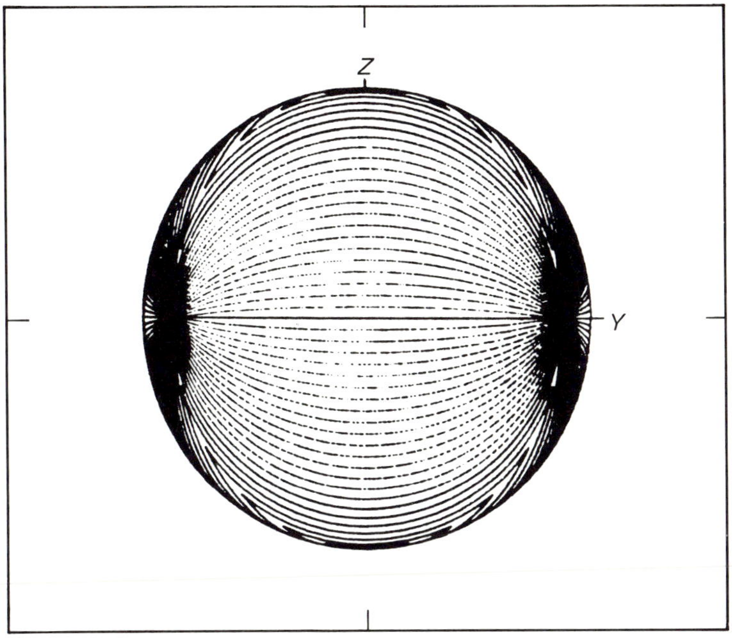

Figure 1 The shape manifold $S^2(\frac{1}{2})$ for the shapes of the labelled triads of points in the plane, carrying contours for the maximum angle of the triangle formed by the points (reproduced, by permission, from ref. 4).

The particular case $k = 3$ is of special importance, and then the shape-manifold is isometric with a $2D$ sphere of radius ½ (this being the appropriate version of $\mathbb{C}P^1$). When k exceeds 3 the situation is just as attractive, save only that the shape-space will then have at least four dimensions and cannot be visualized so easily.

Now the shapes of *exactly* collinear k-ads can be regarded as forming a special set of points lying in the shape-manifold. This set can be identified as a version of real projective space $\mathbb{R}P^{k-2}$. In the important case when $k = 3$, it is a specific great circle on our sphere $S^2(\frac{1}{2})$. We can see it in Figure 1 as the 'equator' running across the centre of the figure. The remaining curves in the diagram show other contours for a function-of-shape which is max (A, B, C), where A, B and C are the angles of the triangle. Thus the set of *nearly* collinear shapes is represented by a union of narrow lenticular regions close to the 'equator' (there are three of these, of which one can be seen in Figure 1, and two partly seen).

So far we have avoided introducing statistical ideas. But suppose we remedy this by agreeing to think of our three points as a sample from a Gaussian (normal) distribution of circular symmetry. This extra degree of specification will provide us with a probability-measure sitting on the shape-manifold, and so will enable us to make probability statements about shapes. In particular, we can compute the probability

$$\text{pr(the triangle } ABC \text{ is } \varepsilon\text{-collinear)}$$

by integrating the above probability-measure over the three lenticular regions corresponding to ε-collinearity. Now the probability-measure in the Gaussian situation turns out to be the uniform distribution over the spherical surface, and so the original problem has been reduced to spherical trigonometry.

In numerical terms a main result (obtained earlier for a similar problem by Broadbent using direct computations) tells us that each such set of 50 points in the plane will turn out to have, among its $\binom{50}{3} = 19{,}600$ triplets, about 130 triplets that are collinear for a value of ε corresponding to 1°. That this result is beyond our intuition is easily tested, by asking any audience, professional or not, to guess the expected number of such nearly collinear triplets. Wherever in the world I put this question, I get answers ranging from small numbers like 2 or 3, up to several thousand.

Of course, you will appreciate that this is only a sketch of what has been done, and that much fine detail has had to be suppressed. Thus we may ask, what happens if we use non-Gaussian distributions, why do we assume circular statistical symmetry, why do we use this value for ε, why do we not consider a range of εs, and so on. All these and many other questions are answered in the paper[3] by W.S. Kendall and myself. The answers are too technical to be appropriate here.

There has been some controversy about the way in which these arguments should be extended to larger values of k. There is no difficulty in describing the geometry of the shape-manifold, and the nature of the shape-measures. What has been in dispute is how we are to assess near-collinearity when there are more than three points. Fortunately the general theory I have been sketching provides a natural and attractive answer. Let us call the subset of *exactly* collinear shapes, 'Coll'. Then Coll is the subset $\mathbb{R}P^{k-2}$ of the shape-manifold $\mathbb{C}P^{k-2}$, and the shape of the k-ad to be tested is a point P lying in $\mathbb{C}P^{k-2}$. The Riemannian geometry provides us with the shortest geodesic distance in $\mathbb{C}P^{k-2}$ from P to Coll, and it is hard to conceive of a more natural measure of non-collinearity than that. If we call the above shortest distance L, then the shape-measure theory in the Gaussian case provides us with the statistical distribution of L, and indeed also with its statistical distribution when the generating Gaussian distribution is 'stretched' ($\sigma_1 = \sigma_2$). This means that (1) we can assess the significance of an observed effect, and (2) we can say how much more likely we are to get such an effect by chance, if the 'stretched' nature of the overall layout is allowed for.

The cosmos provides us with a neat example. Arp and Hazard[5] noticed a curious configuration of six QSOs involving collinearities. As they had quite different individual redshifts, it became important to know whether the observed collinearities could be dismissed as a chance effect. Treatment along the lines suggested here (for a detailed account of which see ref. 4) indicates that the collinearities can at any rate not be dismissed out of hand, the significance level being about 1/400, although it is clear that a firm decision on the matter must await a large-scale survey.

The results of such surveys are now becoming available in work by M.G. Edmunds and G.H. George (Cardiff), and by S.V.M. Clube, A. Savage and A.S. Trew (Edinburgh).

The ideas described above have been extended to the general m-dimensional case, and new features arise when m is 3 or more. The representation-space for shapes then acquires homological torsion, and differential-geometric singularities also appear although for statistical (but not stochastic) purposes there is a 'big' smooth subspace on which most calculations can be carried out. Obvious further generalizations such as (1) the case of infinite k, and (2) the study of the 'shapes' arising when the product of k copies of a homogeneous Riemannian manifold M is quotiented by the action of a transitive group on M, remain problems for the future.

Another recent geometric/statistical study[6] arose as a response to a remarkable paper by Birch[7] noting what he considered might be a large-scale asymmetry of the Universe. His paper has been criticized in three ways by Phinney and Webster,[8] who pointed out that (1) the data could have been more judiciously selected, (2) the statistical significance of the effect was not obvious, (3) the effect, if it exists, could be local rather than global, super-galactically speaking (say as a consequence of incomplete correction for Faraday rotation). We met (1) by working with a revised data-set supplied by Webster, and to investigate (2) we devised what we think is an appropriate and natural significance test. As non-professional astronomers we offer at present no comment on (3). I now describe some of the ideas used in the calculation, which shows that we have here to deal with an observation which, considered on its own, is surprising at a significance level of about 1/2,000. This indicates that Birch indeed made a substantial point, whatever the correct astronomical interpretation may be.

Birch was concerned with 100 or so high luminosity classical double radio sources for which it is possible to determine (*a*) a geometrical elongation axis and (*b*) a polarization axis, and to measure the acute angle Δ between them, this angle being given an algebraic sign according to the direction in which one must move from (*a*) to (*b*) in order to describe that angle. Thus $-90° \leqslant \Delta$ 016 $+ 90°$, and $\Delta = -90°$ must be identified with $\Delta = +90°$, because there is only one way in which the axis (*a*) can be perpendicular to the axis (b). If we think of the sources as living on the celestial sphere (for example, on the surface S^2 of an astronomical globe) then we can write such a single observation in the form (p, Δ), where p is the unit position vector from the centre of the globe (so $p^2 = 1$), and Δ is as above. Evidently, we can think of Δ as a point in $1D$ real projective space $\mathbb{R}P^1$, so that whenever it is convenient we can work instead with 2Δ in S^1. Our set of observations will then be $((p_j, 2\Delta_j) : j = 1, \ldots, n)$ and a statistical analysis must proceed from the postulation of two probability measures for (p, 2Δ), one corresponding to a null Birch effect (Δ independent of p), and the other to a positive Birch effect (Δ statistically linked to p).

It is instructive to examine the topological character of the full sample-space in this problem. A basic data-point is of the form (p, l, m), where the unsensed tangent line l at p locates the geometric axis of the source, and m similarly locates the polarization axis. When p and l are given, then Δ is the angle by which we must rotate l about p to get m. Thus the sample-space for (p, l, m) is a product with the total space T of the $\mathbb{R}P^1$ -bundle over S^2 as first factor (for p and l) and $\mathbb{R}P^1$ as second factor (for Δ). We can replace the second factor by S^1 if we use 2Δ instead of Δ. It will be seen from the Appendix that this full sample-space for (p, l, m) is homeomorphic to $L(4, 1) \times S^1$, where $L(4, 1)$ is a particular one of the so-called lens spaces. The calculations which follow take place in the marginal space for the reduced observations (p, 2Δ), and so require only the reduced sample-space $S^2 \times S^1$, but in further work it is likely that the lens space will play an important role.

The effect described by Birch amounts to a topographic regression of Δ (or 2Δ) on p, and in everyday terms it can loosely be described as follows. It is as if, when observing the heavens in one hemisphere of directions-of-viewing, we see a large number of capital roman letters N, together with a few capital cyrillic letters И, while in the opposite hemisphere of directions-of-viewing there are many cyrillic Иs and a few roman Ns. (Here the Ns correspond to say $0 < \Delta < \frac{1}{2}\pi$, and then the cyrillic Иs correspond to $-\frac{1}{2}\pi < \Delta < 0$; of course, to make the simile more appropriate we must envisage a mixture of Ns, cyrillic Иs, and 'intermediate' letter-shapes.) Such a phenomenon would be observed if the objects in question were maximally distant from us in Dante's S^3-universe, so that what we see in one direction as N, we also see again (from behind) in the 'opposite' direction as cyrillic И. This was not Birch's interpretation, of course, but it might have been the one that would have been put forward 50 years ago.

As p specifies a point on the celestial sphere S^2, it represents a direction in the ordinary sense. As Δ (to be coded in the calculations as 2Δ) specifies a point on the projective space $\mathbb{R}P^1$, we speak of it as an 'indirection'. The terminology here is borrowed from Shakespeare, and what he makes Polonius say is in fact remarkably appropriate to the present investigation:

"And thus doe we of wisedome and of reach
With windlesses, and with assaies of Bias,
By indirections finde directions out."

We are thus concerned with a regression of an indirection on a direction.

While dealing with literary matters I had better add a few words on 'Dante's S^3-universe'. That Dante correctly described S^3, and perhaps seriously thought of it as a model for the universe, was pointed out by Peterson,[9] who assembled evidence from the *Paradiso* indicating a representation of the Universe as the union of two copies of a solid sphere B^3. The first has the Earth at its centre, surrounded by the concentric shells of the Aristotelian universe, and ultimately by the Primum Mobile as its boundary S^2. The second copy of B^3 has a similar structure of concentric spherical shells, this time with God at its centre, and again the Primum Mobile as an S^2-boundary. It seems that Dante intends us to view the whole Universe as composed of these two copies of B^3 joined together by a 1:1 glueing together of their two boundaries S^2. And this is, of course, the standard construction of S^3. I am indebted to Professor Manfred Gordon for this information.

At the heart of any mathematical formulation for a regression of the kind we have in mind, one will expect to find an expression of the form

$$\alpha \cos 2\Delta + \beta(q.p) \sin 2\Delta,$$

where p is the observed direction, q the directional pole of the regression, Δ records the observed indirection, α controls the null effect, and β is the regression coefficient. It is natural to build up the model by making use of the Fisher and von Mises distributions familiar in directional statistics (for which see ref. 10), and, of course, we must incorporate a factor $f(p)\mathrm{d}p$ to describe the distribution of source-positions p on S^2. It would be inappropriate to postulate any simple form for the function $f(p)$, which describes the sociology of observatory-building rather than an astronomical phenomenon. These considerations led us to write

$$\frac{\exp(\alpha \cos 2\Delta + (\lambda\cdot p) \sin 2\Delta)}{2\pi I_0(\surd(\alpha^2 + (\lambda\cdot p)^2))} \; f(p)\, \mathrm{d}p\, d\,(2\Delta)$$

for the joint distribution of p and 2Δ, where now we have collapsed the parameters β and q into the vector $\lambda = \beta q$. We need to test the null hypothesis $|\lambda| = 0$ against the alternative hypothesis $|\lambda| > 0$, in the presence of the nuisance parameters α and q. ($I_0(\cdot)$ in (1) is the zero-order modified Bessel function of the first kind.)

The standard procedure in such cases is to multiply together the factors at (1) for each observed pair (p_j, $2\Delta_j$) so as to form what is known as the likelihood of the observations, and then to maximize this as L_1 when all parameters are free, and as L_0 when $\beta = |\lambda|$ is forced to zero. The conventional Neyman–Pearson test statistic is then $U = 2 \ln(L_1/L_0)$, and it is important to notice that there will be no contributions from $(f(p_j) : j = 1, \ldots, n)$ to U, so that we are not committed to any speculation about the form of the function f. It is known that asymptotically, when n is large, then U will have approximately the chi-square distribution with 3 degrees of freedom, but one would be hesitant to rely too much on the relevance of that fact in the present instance, where n is only 134.

The maximized value L_0 is easily found. The calculation of L_1 is more difficult as it involves a global search over a four-dimensional space, but as it happens this is simplified by convexity considerations. It thus turns out that $U = 16.04$. The asymptotic arguments mentioned above indicate significance at a level of about 1/1,000, but we do not wish to proceed in that way with such a small number of sources, so instead we use what is called a 'data-based' simulation test.

The idea behind this is to keep the partitioned sets of observations (p_1, $p_2, \ldots, p_n$) and ($2\Delta_1$, $2\Delta_2, \ldots, 2\Delta_n$) unchanged, but to pair them up in random order, so as to erode the effects of any real regression that may be present. After each such randomization we compute the corresponding U-value, carrying out this operation some conveniently large number of times, and we then compare the observed $U = 16.04$ with the large collection of simulated test-statistics which are taken to represent the U-distribution that would have been obtained in similar circumstances if, in fact, there were no real Birch effect. The time-consuming global optimization involved in the calculation of L_1 makes it very costly to perform more than a few hundred such simulations, and so we use the fact that it is much quicker to calculate a first-order approximation V to U. The value of V for the actual data is $V = 14.64$, and this, when judged against 10,000 simulated V-values, turns out to be significant at a level of about 1/2,000. Similar calculations with U, necessarily smaller in extent, confirm this result.

Accordingly we are driven to conclude that there is a strong indication of a real Birch effect which calls for explanation. Its estimated pole is

$$RA\ 13^{\mathrm{h}}30^{\mathrm{m}},\ \mathrm{Dec.}\ -37.4^\circ,$$

with an uncertainty of about 20°–30°, so that, to be acceptable, an explanation should imply the existence of a pole within that distance of the estimated pole.

In thanking you again for inviting me to deliver this lecture, and IBM UK Ltd for its generous patronage, I should like to say how pleasant it always is to visit the Mathematical Institute of which Milne was for many years the Curator.

References

1. von Mises, R., 'Uber die 'Ganzzahligkeit' der Atomgewichte und verwandte Fragen', *Phys. Z.*, *19*, 490–500 (1918).
2. Fisher, R.A., Dispersion on a sphere, *Proc. R. Soc. A*, *217*, 295–305 (1953).
3. Kendall, D.G., and Kendall, W.S., 'Alignments in two-dimensional random sets of points', *Adv. appl. Probab.*, *12*, 380–424 (1980).
4. Kendall, D.G., Shape manifolds, procrustean metrics, and complex projective spaces, *Bull. Lond. Math. Soc.*, *16*, 81–121 (1984).
5. Arp, H., and Hazard, C., 'Peculiar configurations of quasars in two adjacent areas of the sky', *Astrophys. J.*, *240*, 726–736 (1980).
6. Kendall, D.G., and Young, G.A., Indirectional statistics, and the significance of an asymmetry discovered by Birch, *Mon. Not. R. astr. Soc.*, *207*, 637–647 (1984).
7. Birch, P., 'Is the universe rotating?', *Nature*, *298*, 451–454 (1982).
8. Phinney, E.S. and Webster, R.L., 'Is there evidence for universal rotation?', *Nature*, *301*, 735–736 (1983).
9. Peterson, M.A., 'Dante and the 3-sphere', *Am. J. Phys.*, *47*, 1031–1035 (1979).
10. Mardia, K.V., *Statistics of Directional Data*, Academic Press, New York (1972).
11. Rolfsen, D., *Knots and Links*, Publish or Perish Inc., Berkeley (1976).

Appendix. The total space T of the $\mathbb{R}P^1$-bundle over S^2

A typical element t of T can be expressed in the form (p, l), where as usual p is a unit vector locating a point on S^2, and l is an 'unsensed' tangent *line* for S^2 at that point. Suppose that $l = (\lambda v : \lambda \text{ real})$ where v is a unit tangent *vector*; of course l is not changed if we replace v by $-v$. We can think of p and v in that order as the first two members of a right-handed mutually perpendicular triad (p, v, w) of unit vectors got from a basic such triad $(\mathbf{i}, \mathbf{j}, \mathbf{k})$ by the quaternionic operation

$$(p, v, w) \to q\,(\mathbf{i}, \mathbf{j}, \mathbf{k})\,q^{-1}$$

where q is a unit quaternion $x_0 + x_1\,\mathbf{i} + x_2\,\mathbf{j} + x_3\,\mathbf{k}$ $(\Sigma x^2 = 1)$. We can identify $q = (x_0, x_1, x_2, x_3)$ as a point of S^3, and we can also think of $q = (x_0 + ix_1, x_2 + ix_3)$ as a point of $\mathrm{C} \times \mathrm{C}$. Both interpretations will be useful here.

It is clear that we must identify q^* with q whenever both

$$q^*\,\mathbf{i}\,q^{*-1} = q\,\mathbf{i}\,q^{-1} \text{ and } q^*\,\mathbf{j}\,q^{*-1} = \pm\, q\,\mathbf{j}\,q^{-1},$$

and in terms of $Q = q^{-1}q^*$ these relations are equivalent to

$$Q\,\mathbf{i} = \mathbf{i}\,Q \text{ and } Q\,\mathbf{j} = \pm\,\mathbf{j}\,Q$$

so they assert exactly that $Q = y_0 + y_1\,\mathbf{i} + y_2\,\mathbf{j} + y_3\,\mathbf{k}$ satisfies

$$y_2 = y_3 = 0,$$

and

exactly one of y_0 and y_1 is 0.

In terms of the original quaternion q, this just means that we are not to distinguish between q, $\mathbf{i}q$, $-q$, $-\mathbf{i}q$, and thus $T = S^3/(1, \mathbf{i}, -1, -\mathbf{i})$. A moment's algebra shows that when q is considered as a point of C × C then the system of identifications is exactly such that each (z_0, z_1) as a point of C × C is to be identified with (iz_0, iz_1). But this last prescription is one of the several definitions of the lens space $L(4, 1)$ given in ref. 11.

Another model of $L(4, 1)$, also described by Rolfsen, starts with the ball B^3 in R^3 with its boundary S^2 split into eight closed abutting pieces, these being then sewn together as follows. From the upper hemisphere we obtain four quadrants H_0, H_1, H_2 and H_3, successively related to one another by rotation about the positive z-axis through 90°. From the lower hemisphere in a similar way we obtain H'_0, H'_1, H'_2 and H'_3, but we arrange for H'_0 to lie directly below H_1, H'_1 below H_2, H'_2 below H_3 and H'_3 below H_0. The sewing programme is then: for each j, H_j is to be sewn to H'_j in a manner respecting the 90° rotations about $OZ+$ and the reflections in the plane $Z = 0$. What results from this operation is a closed connected (but not simply connected) orientable 3-manifold which is homeomorphic to $L(4, 1)$ as previously introduced, and so to T.

These representations show that the integral homology of T is $Z/4$, 0, and Z in dimensions 1, 2 and 3, and that its fundamental group is $Z/4$.

I am indebted to Professor J.F. Adams, FRS for assistance with this Appendix.

Dr. Geoff W. Robinson

Geoff Robinson is Manager of the IBM UK Scientific Centre at Winchester. He obtained a first degree in Mathematics at the University of Nottingham in 1966. His doctorate was awarded by the University in 1969. His PhD thesis was concerned with Quantum Chemistry calculations of protonation effects on aromatic bases.

He joined IBM at the Hursley laboratory in 1969. His computing experience with the company includes languages and compilers, data base and data communications systems, office systems, and personal workstations.

He took up his present appointment in 1982. His particular goal at the Centre is to promote an environment of co-operation between IBM staff and external partners in the examination of the use of various new technologies to improve human interactions with computers.

Perception from pictures

In IBM, research and development is divided into separate parts of the company. It can be argued that the task of the research people is to convert uncertainty into probability, in terms of technology, applications, business opportunities and so on. The development people then have to convert this probability into certainties from which the company is going to make some money. The scientific centres form part of IBM's research, rather than product development activities.

There are 16 scientific centres in IBM located around the world from Cambridge, Massachusetts to Cairo, from Mexico City to Madrid. They are fairly small organizations. Their staffs range from about half a dozen people in Caracas, to 40 to 50 people in Rome. They are very cohesive organizations and work freely with people outside IBM, which creates some unique challenges. IBM staff are on the whole computer people and not always leading scientists in specific disciplines. So IBM and non-IBM staff mix in order to pursue scientific, engineering, medical, technical research with a heavy computer content. Each centre tends to have its own specialization. Joe Raviv in Israel, for example, does much work on mathematical modelling. In Winchester, the emphasis is on communicating the information and data in computers to scientists, clinicians or whoever is sitting in front of the computer screen. In particular, there are research groups working on image processing, computer graphics and speech synthesis.

Image processing

Much of the image processing work is centred on medical images. Figure 1 shows a poor quality Victorian photograph of a minor in County Durham. It is the first medical image in the sense that it was created by technology.

Figure 2 shows another example of a medical image, a standard chest X-ray. One project at the Winchester centre is concerned with enhancement techniques applied to such images. This part of IBM's work is done with Professor David Denison and the radiology group of the Brompton Hospital in London. This standard chest X-ray shows the shadow of the heart,

Figure 1 Victorian photograph of a coal miner.

Figure 2 A standard chest X-ray.

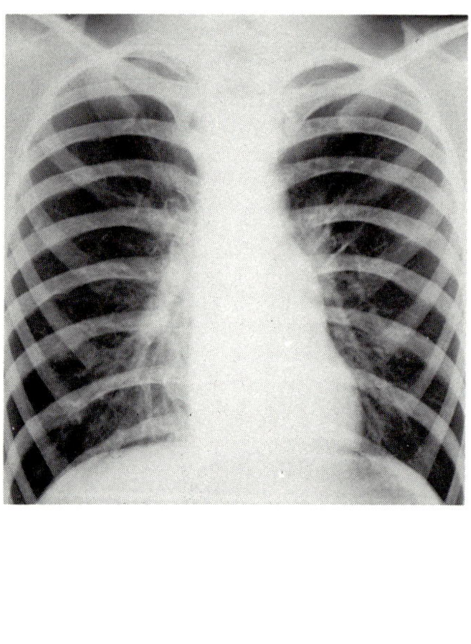

Figure 3 A standard chest X-ray showing the application of simple image-processing techniques.

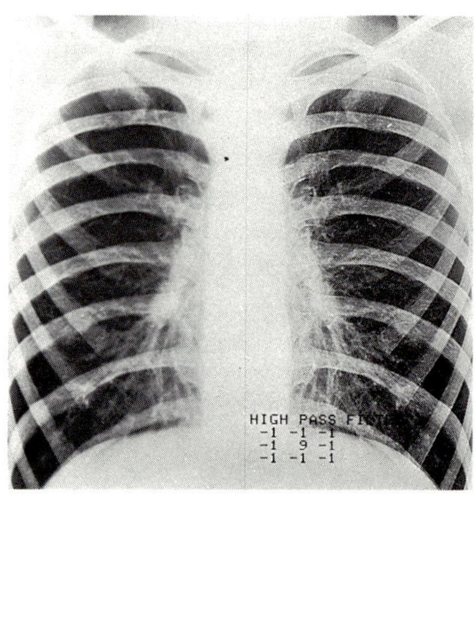

the ribs in the various orientations at the front and back, and the fine structure which is the blood vessels in the lung. By applying simple image processing techniques, as in Figure 3, a three by three convolution on this image, the structure of the blood vessels in the lung, can be heightened.

Simple techniques such as the above have clinical value. Figure 4 shows local grey-level remapping which has the effect of shining black light through a medical image, just by darkening parts of the image. At the top it enhances the vertebrae, and the ribs can be seen behind the heart. Note that all information was actually on the original medical image, the original X-ray photograph, it is just that the human eye and the human brain cannot distinguish it easily in that form.

Figure 5 has more clinical significance. It shows a pneumothorax which occurs when the lung wall falls in from the wall of the chest. The lung is normally held out by pressure. If the lung is punctured then the wall of the lung will collapse inwards. In such a case, the edge of the lung is very difficult to detect on an X-ray. Figure 5 shows both the original and enhanced images of a pneumothorax. The lung wall is the faint white line running down the right-hand side. All the technique does basically is stretch out the chest wall into a square, shift it slightly and subtract it. This highlights the radial edges.

Figure 4 Grey level remapping of a medical image.

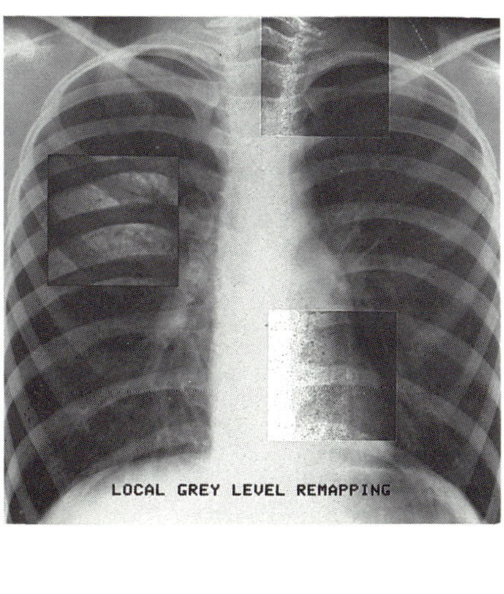

Figure 5 Original and enhanced images of a pneumothorax.

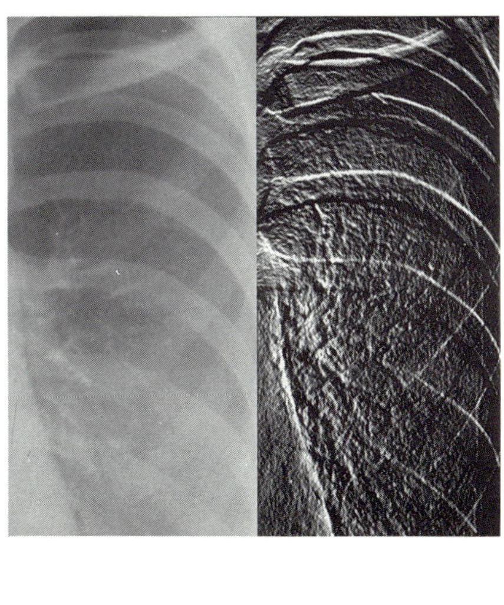

A different application, although still in the X-ray field, being researched in conjunction with the Queen Alexandra Hospital at Cosham, is for orthodontic X-rays. These show the side of the skull and the jaw. They are taken in order to examine facial deformities or misaligned teeth. Traditionally, orthodontists take these X-rays and trace them onto tracing paper in order to identify certain significant points on the image. From these, distances and angles can be measured to obtain the underlying geometry of the skull. By providing the orthodontists with various enhancements on the computer screen, it is easier for them to identify the particular points they are looking for, and let the computer provide the resultant geometry. Both the hard edges of the skeletal structure, and the soft edges of the face itself can be detected. In practice, as Figure 6 shows, they have a variety of different enhancements to choose from. As in all this work, it is not the computer sophistication that is important, but the value to the clinicians looking at the pictures.

Another project, being carried out in collaboration with Professor Lorber at Sheffield, is the study of brain ventricles. Figure 7 shows a normal scan of the human brain as a series of horizontal slices. The interesting points are the small black areas in the centre, which are the brain ventricles. These are small bodies in the centre of the brain, where cerebro-spinal fluid is

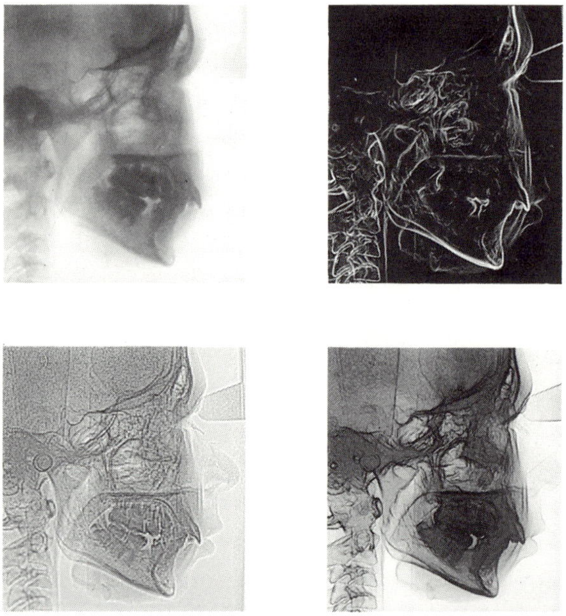

Figure 6
Orthodontic images.

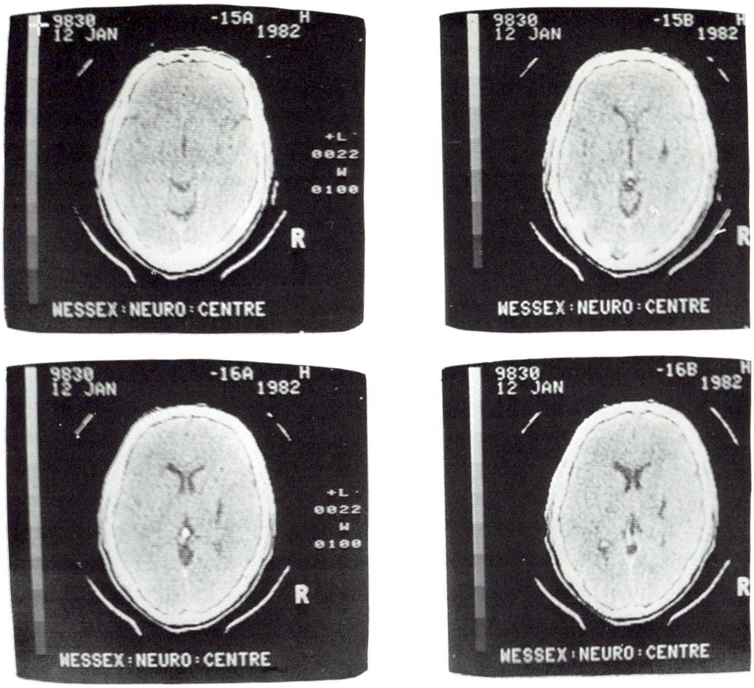

Figure 7 Normal scan of the human brain as a series of horizontal slices.

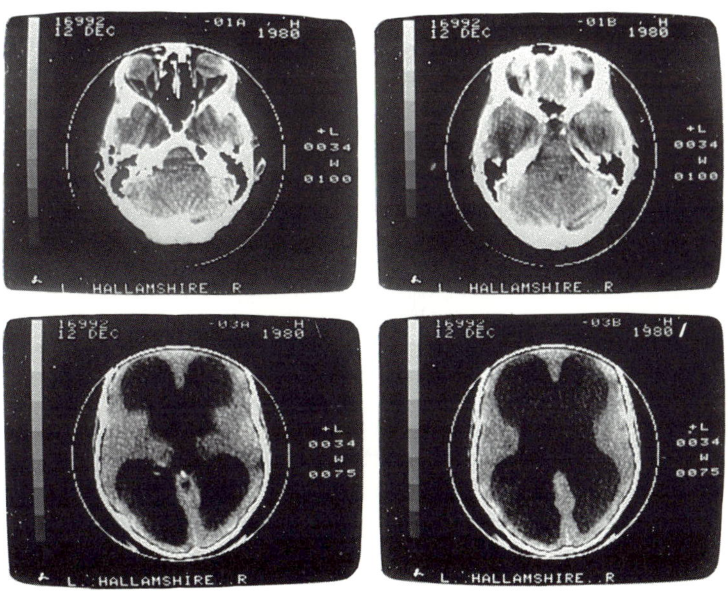

Figure 8 An example of extreme hydrocephalus.

generated and recirculated. In the condition known as water on the brain or hydrocephalus, the outlets of these ventricles get blocked and the ventricles enlarge. Figure 8 shows an example of extreme hydrocephalus, with the ventricles apparently occupying a major part of the skull. It would be interesting to reconstruct a three-dimensional model of the brain and the ventricles so that their volume and their exact shape can be calculated. Calculations show that slightly more than 50% of the volume of this brain is taken up by these enlarged ventricles. Surprisingly this patient has nothing clinically wrong with him, in fact, he has a first class honours degree.

A slightly different type of image is shown in Figure 9. This is a single image from a coronary angiogram. The clinician uses this technique to study how the left ventricle of the heart is functioning, this being the part that pumps the blood to most of the body. Most heart attacks start with arhythmia in the left ventricle. What one would like to do before surgery is to study the operation of the left ventricle; to do that one needs to take a cine X-ray film of the left ventricle. The problem is that the heart does not show up on an X-ray, so to overcome this, a catheter is inserted via the wrist and arm into the heart and a radio-paque dye is then injected directly into the left ventricle so that the heart will show up on the X-ray film. These techniques are commercially available using machines that digitally subtract the two images before and after the dye is injected and produce a fairly clear picture of the left ventricle itself. In the real application this would produce an X-ray film that could be watched on a TV monitor. The work at Winchester in this area is devising techniques for producing three-dimensional models of the coronary arteries.

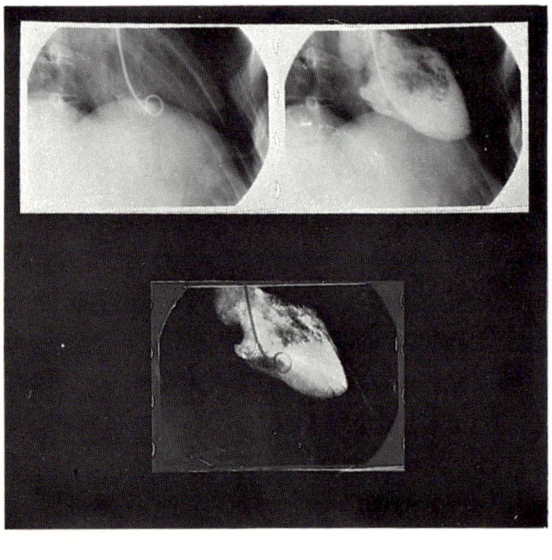

Figure 9 Single image from a coronary angiogram.

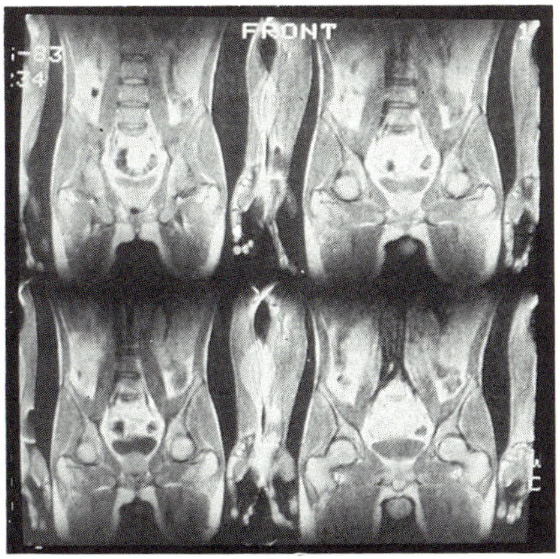

Figure 10 A nuclear magnetic resonance image of the thorax of a human patient.

Figure 11 The sublety of human perception; this shows a picture of a cow which is not helped by computer enhancement.

Figure 10 shows a nuclear magnetic resonance medical image which produces very clear images of the inside of the human body. This exciting new technology is just starting to take off. In other work, in progress with the co-operation of David Denison's group at the Brompton Hospital, the method used by the human patient to breathe is being investigated. Light is shone onto the thorax of the patient as he is breathing; the light is very carefully defined patterns on standard 35mm slides which are projected onto the patient's body. Pictures are taken from which a digitized position is obtained and a three-dimensional view of the exact shape of the patient's body while he is breathing can be constructed. The area where this technique could have the greatest potential application is for the tetraplegic. It provides a very accurate way of measuring what is actually happening to a totally paralysed patient's breathing.

Perception

In image processing the main problem is one of perception. Figure 11 illustrates the sublety of human perception. Without help, most people cannot identify it. It is, in fact, a picture of a cow. Computer enhancements of this image do not aid our recognition of it. Recognition is difficult because many of the high-frequency features—the sharp black and white edges—are misleading; computer enhancement tends to emphasize these features. Recognition is only possible in the human brain by some process, not understood, of semantic comparison with a 'stored library' of objects.

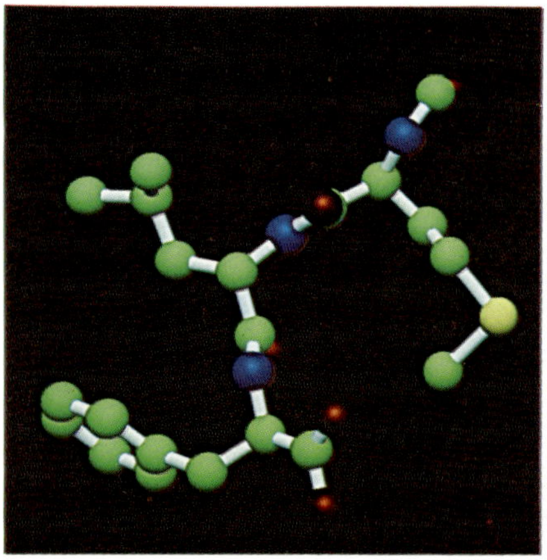

Figure 12 A conventional ball and stick model of the molecule formyl methionyl leucile phenylalanine.

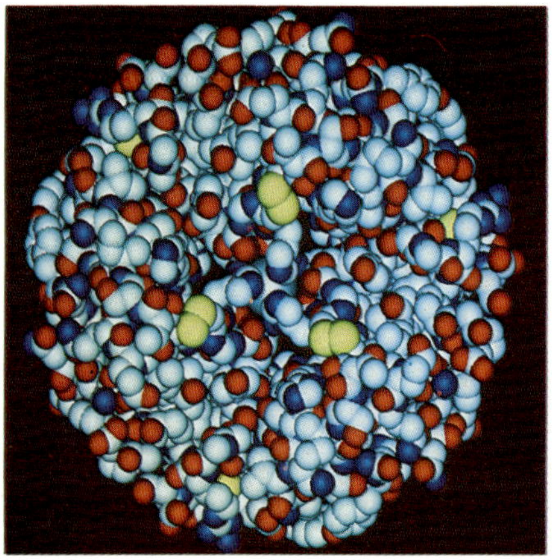

Figure 13 A representation of a hexamer of insulin.

Computer graphics

Live objects are made up of small molecules, and computer graphics can be used to improve the understanding that research chemists, biochemists, pharmacists and related specialists have of how biological processes take place. Many of these processes involve the interaction between small molecules and proteins. There are various ways of representing such molecules. Figure 12 is a conventional ball and stick model of formyl methionyl leucile phenylalanine, a small drug-like molecule. The problem with this technique is that for larger molecules, such as proteins, such representations become confusing. Figure 13 shows a representation of a hexamer of insulin. Like all protein molecules, insulin is made up of long sequences of amino acids folded into a three-dimensional structure. One problem for crystalographers has been, given an electron density map of the three-dimensional shape of a protein molecule, and given, from chemical experiments, the sequence of amino acids, then how did that sequence of amino acids fold up into the three dimensional shape. There are techniques which can help to make the representation of the structure clearer.

Figure 14 shows a representation of the lysozyme protein in work done in collaboration with Arthur Lesk at the Medical Research Council at Cambridge. The ribbon effect of the amino acid sequence is shown by the continuous line. In certain parts, this line goes into a cylindrical shape; these cylinders represent the alpha helices in the protein. This line and cylinder combination can provide a simple view of the global structure of

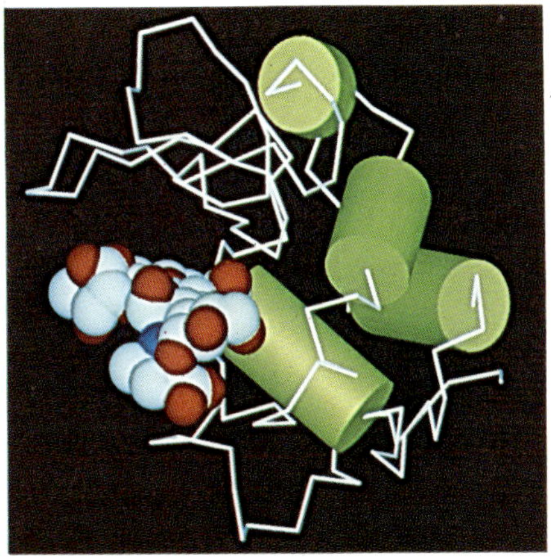

Figure 14 A representation of the lysozyme protein.

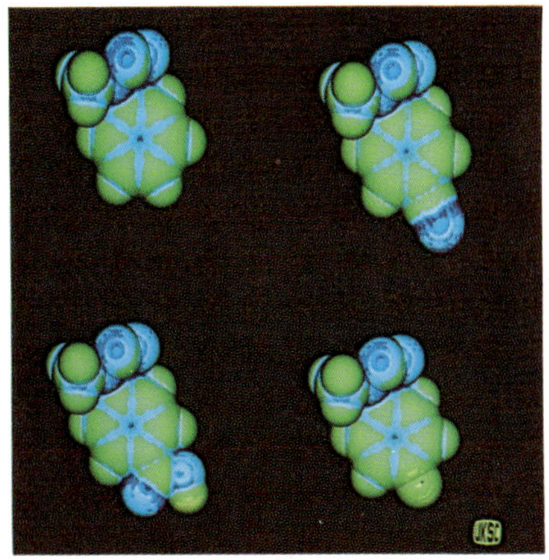

Figure 15 Part of a drug sequence.

the protein. All protein molecules have an active site, which is where the small activator or inhibitor docks with the protein. Figure 14 also shows such a substrate molecule docking in the active site of the lysozyme protein. This technique gives a first test as to whether a drug molecule will actually fit into that active site or whether it will be barred. Unfortunately whether it will fit or not is only one part of the problem. There are other things that one has to worry about, for example, the electrostatic charge distribution. The two molecules are not going to come together if Coulomb forces drive them apart. The results of quantum mechanical calculations of the electrostatic potential over the surface of small drug-like molecules can be represented as a colour-coded surface. In the design of drugs it is of interest how this potential distribution is altered, as alterations to the drug molecule are made. In a sequence of derivatives of the same molecule, the electrostatic potential distribution may alter significantly as the substituent changes. A pictorial format as opposed to pages of numbers of the electrostatic charge distribution immediately gives an impression of the activity in any particular form of the sequence of drugs. Figure 15 shows such a sequence.

Computer graphics provides a powerful tool for examining data in a wide range of scientific disciplines. IBM has taken an interest in archaeology in a project with the Museums in Southampton, investigating a Saxon bone pit in the city of Southampton where the archaeologists have an interesting data problem. The bone pit is about a metre and a half in diameter, about a metre and a half deep, and each bone found in this pit

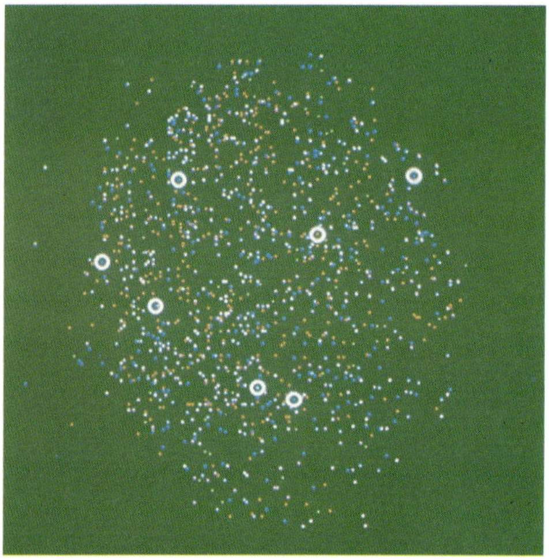

Figure 16
Highlighting and identifying bones within a stratum.

has been meticulously measured with regard to its position in three-dimensional space, its orientation, its species type, the markings that were on it and so on. These data can only be of value if they can readily be used to answer questions of archaeological importance. Which species were close together? Did the species change with time? Did the types of marking on them change with time, indicating whether butchers were using it as a pit, and so on. By giving the archaeologists access to our graphics system which includes a data base to hold all their information, they have a powerful way of asking these questions and a pictorial way of understanding the answers. By colour coding the bone fragments, the computer system could highlight, for example, all cow bones and identify them within a particular strata (Figure 16). From then on archaeologists can identify a particular bone, and then look for all the other cow bones within a certain radius of that cow bone, not only in that strata but above and below it as well. The archaeologists thus have a powerful tool to investigate large amounts of data when there is no other scientific way of investigating the implications of it.

Conclusion

In the wide range of studies carried out at the Winchester Scientific Centre, there is a common theme. How can the power of the computer, particularly its ability to generate, store and manipulate pictures, assist the scientist in his exploration of the unknown, or the uncertain.

Mr. Jim Baldwin

Jim Baldwin is Reader in the Department of Engineering Mathematics, and Director of the Information Technology Research Centre at the University of Bristol. After graduating from Leicester University, he worked with Hawker Siddeley researching system theory for missile modelling, and lectured in Mathematics at Hatfield Polytechnic.

His research interests include Fuzzy Logic, artificial intelligence, decision support systems, and the general area of information technology. The recently formed Information Technology Research Centre includes contributions from a number of disciplines—computer scientists, an AI group, a microelectronics group, the MRC Brain and Perception Laboratory, psychologists and philosophers.

Fuzzy reasoning: the need for blunter tools

It has been recognized that human reasoning is difficult to model mathematically. With the advent of EXPERT SYSTEMS this form of reasoning is required by modern computer networks to provide the computer with a true emulation of man's expertise.

The essential vagueness of human thought was described by Russell,[1] in the course of which he remarked that classical logic appears to be "not applicable to this terrestrial life but only to an imagined celestial existence". On a more pragmatic note, Bellman[2] says

> "For most ideas, mathematics is far too simple a language. We require the medium of ordinary language with its richer content, and therefore greater inherent capacity for ambiguity. That ambiguity is required for meaning, that precision requires imprecision are paradoxes . . . the human mind, remarkably and fortunately, can cope with ambiguity."

The application of fuzzy logic to the resolution of certain paradoxes has been discussed by Baldwin and Guild[3], the discussion being based on an approach to fuzzy logic given by Baldwin[4] which relies on the theory of fuzzy sets given by Zadeh.[5] Human intelligence uses a much less precise model of reality replacing microscopic precisional description with a more macroscopic fuzzy one. The chess problem is an example where the complexity arises because of the high dimensionality and the large number of decisions which are possible at each stage of the game. The rules and goals of the game are well defined but it is this microscopic precision which prevents a mathematical solution being obtained since it gives rise to a computationally impossible analysis. For the fuzzy approach a different state space is used (one of lower dimension) and the introduction of this simplified description taking into account only what is really relevant, allows human intelligence to gain insights into a chess problem resulting in a high standard of decision making, even though the decision policies will necessarily be fuzzy in nature. The analysis is in terms of board patterns rather that exact positions of each piece on the board.

For other problems, for example, political decision making, economic planning, domestic and other human problem solving, the state mappings from stage to stage will only be known in an imprecise manner. We all feel that we are able to choose the best car route when going on holiday and this is arrived at using a very fuzzy decision analysis. Natural language understanding requires fuzzy logic. Consider, for example, the following attempt to define what is meant by an humane industrial society.

An humane industrial society is one for which EVERYONE has at least a REASONABLE standard of living with a FAIR distribution of INCOME where ALL people have the opportunity of SATISFYING and INTERESTING work and pastimes. It should be FLEXIBLE but STABLE, both industrially and socially, and be ADAPTABLE to outside influences. EQUAL educational opportunities should be available of HIGH standard, GOOD medical facilities provided, and the DIGNITY of man RESPECTED. Each of the words in capitals represented a fuzzy concept, even ALL means MOST. Fuzzy concepts cannot be given a precise algorithmic definition. Examples of the concept can be given and these examples will share a family resemblance in the Wittgensteiniun sense. Fuzziness results from a lack of definition rather than current state of knowledge. Trees resemble each other in a variety of ways without there necessarily being any one resemblance that exists between any pair. The border of applicability of a predicate is fuzzy. Between applicable and not-applicable there is a degree of applicability. Fuzzy qualifiers such as MOST, SEVERAL, MANY, FEW, and fuzzy truth values such as QUITE TRUE, VERY TRUE, MOSTLY FALSE and fuzzy probabilities exemplified by LIKELY, UNLIKELY, etc. abound in natural language. Common sense reasoning employs such fuzzy terms.

Fuzzy linguistic statements can provide an economical representation of data. This summary can then be used with similar summaries from other sources for modelling and fuzzy reasoning. One must balance the need for precision with the need for general perception. A photograph of too large a landscape gives only a very general impression, while a snapshot of a face tells much about the face but nothing of the surroundings. A principle of uncertainty is applicable to all man/machine systems in which to know something about the system is to necesarily interact with that system. If the system is observed from afar with only weak interaction, then only very vague and fuzzy knowledge will be obtained. As the observer strengthens his interaction with the system, this interaction becomes more concentrated about a smaller and smaller part of the system, so learning more and more about less and less.

A principle of uncertainty of databases exists which says:

(1) more detail means more specialization leading to tunnel vision;
(2) summaries of background knowledge are required.

Examples are

Most American cars are large.

Most tall people who are not thin are heavy.

Few bone cancer patients recover.

This background knowledge can be used for reasoning purposes as a default in the absence of more specific information. New knowledge will only be inserted into the knowledge base if it is an exception with respect to the expected knowledge using the default data base.

The motivation for introducing a fuzzy aspect to logic and set theory is to help model complex systems and provide a reasoning mechanism more akin to human reasoning for solving problems and taking decisions. It recognizes that there is more than one type of uncertainty and tries to quantify vagueness associated with lack of definition. This should allow a better understanding of natural language and provide a richer language for modelling reality than classical mathematics.

A fuzzy relational inference language

A computer language, a fuzzy relational inference language (FRIL) is now used to illustrate how fuzzy uncertainty can be modelled and used for modelling human knowledge. This has already been described in detail,[6] and only the basic ideas will be discussed below.

FRIL is a high level language based on the mathematics of fuzzy relations incorporating the ability to represent both fuzzy and probabilistic uncertainties. The design of FRIL includes many ideas on fuzzy logic and automated inference by Baldwin and Zaden.[7] It has also been influenced by the work on Codd's relational data bases as described by Ullman[8] and logic programming as given by Kowalski.[9] It is a general programming language which can be used for artificial intelligence work, designing expert systems, decision support systems, intelligent knowledge bases, fuzzy systems theory, etc.

The inference mechanism of FRIL is more akin to a relational data base, but the method of programming using the language is similar to Horn clause logic programming with a restricted unification. This restriction is not necessary and is removed in the next version of FRIL. The more general version will be designed to exploit parallel processing and emphasize streams and processes but the essential nature of the inference mechanism will be unchanged.

Knowledge base of FRIL

A programme in FRIL in its simplest form is a knowledge base containing facts and rules. Queries are then asked by the user and the inference mechanism returns the solutions.

Consider the following knowledge base

(a)

Persons	Name	Ht	Wt	χ
	Jon	6 – 0	12 – 0	1.0
	Harry	5 – 7	11 – 7	1.0
	Jack	5 – 9	13 – 0	1.0
	Mary	5 – 8	11 – 5	1.0
	Jill	5 – 4	13 – 0	1.0
	Fred	5 – 6	9 – 2	1.0

Likes	Name	Name	χ
	Jon	Jill	0.7
	Jon	Mary	1.0
	Jack	Fred	0.6
	Fred	Harry	0.5
	Harry	Mary	0.2
	Mary	Jon	0.7

Tall	Ht	χ
I-type	5 – 10	0
	6 – 0	1.0
	7 – 0	1.0

Heavy	Wt	χ
I-type	10 – 0	0
	13 – 0	1.0

Very	χ=value	χ
I-type	0.5	0
	1.0	1.0

(b)

FRIENDS $(x,y) \leftarrow$ LIKES (x,y) & LIKES (y,x)

GOOD_FRIENDS $(x,y) \leftarrow$ VERY FRIENDS (x,y)

SHORT_PERSON (x) $\leftarrow$ PERSONS $(x,y,\_)$ & NOT TALL (y)

FAT_PERSON $(x) \leftarrow$ SHORT_PERSON(x) & PERSONS (x,u,y) & HEAVY (y)

POSS_ATHLETE $\leftarrow$ PERSONS $(x,y,\_)$ & TALL (y)

HAS_GOOD_FRIENDS $(x) \leftarrow$ FRIENDS (x,y) & POSS_ATHLETE (y)

Facts are given in (a) as relations. For example, the relation PERSONS has three attributes, namely, NAME, HT, WT and a χ column. Associated with each attribute is a domain of possible values and this domain can take the form of a set, a numerical interval or objects of a general language. The χ should not be interpreted as an attribute but a special uncertainty measure which can be given various interpretations. The relation consists of tuples each having a χ value representing the degree to which that particular tuple satisfies the relation.

The tuple (JON 6 – 0 12 – 0) satisfies the relation PERSONS at the level 1. This is equivalent to the logic statement

'JON is a PERSON with HT 6 – 0 and WT 12 – 0'

If we consider the LIKES relation then the tuple

(JON JILL)

satisfies the relation LIKES at the level 0.7. This means that the truth that JON likes JILL is 0.7. Alternatively one can treat this truth as a compatibility. The PERSONS and LIKES relations are called base relations while the remaining tables in the knowledge base are called I-type relations signifying that interpolation is allowed. For example, the relation TALL represents a fuzzy set on the height space. The height of 6 – 0 corresponds exactly to the concept of tall and 5 – 10 has no correspondence with tall. Heights between 5 – 10 and 6 – 0 have a correspondence measure with tall which lies in the interval [0,1], the actual value determined using linear interpolation.

Thus, in general, each tuple of a relation has a χ value associated with it which represents the degree to which that tuple satisfies the relation.

The (b) part of the knowledge base consists of rules of the form of Horn clauses. For example, the first rule states that

For all x and y
x and y are friends if x likes y and y likes x

where x and y are variables and can take any value from the NAME domain. A short person is defined as someone who has a height y and some weight and y corresponds to a height which does not satisfy the tall relation. Relations can be concatenated, as for example, VERY FRIENDS in the above rules. This corresponds to truth functional modification in the literature. For a given tuple, the χ value of the FRIENDS relation becomes the argument of the CHI-value attribute of the VERY relation so that the CHI-value of the VERY FRIENDS relation is then the corresponding χ value of the VERY relation with this argument of the CHI-value attribute. The relation NOT TALL is another example of the concatenation of two relations and is treated in the same way with the following definition of NOT

Not	CHI-value	χ
I-type	1.0	0
	0	1.0

Queries

The query

'Who is a tall person?'

is asked by entering the query

WHICH(*x* PERSON (*x*,*y*,_) & TALL(*y*))

which is translated by the system to

PROJ((1); JOIN(((2,1)); PERSON, TALL))

and then evaluated to give the solution

Solution	Name	χ
	Jon	1.0
	Mary	0.5

The compiled code PROJ((1); JOIN((2,1)); PERSON, TALL)) means that the relations PERSON and TALL are joined and the resulting relation projected onto the first column. The join is computed by taking any tuple from the PERSON relation and any tuple from the TALL relation and accepting the join of the two tuples if the value in the second column of the PERSON relation matches the value in the first column of the TALL relation. If the join is accepted the χ value associated with the resulting tuple is the minimum value of the χ values of the two tuples. The PROJ is evaluated by removing all attribute columns from the resulting relation except the first and removing any duplicates. If duplicates do occur the accepted tuple is that associated with the maximum χ value. This corresponds to using a fuzzy logic in which

$$v(A \text{ and } B) = \min(v(A),v(B))$$
$$v(A \text{ or } B) = \max(v(A),v(B))$$

where $v(X)$ represents the truth of proposition X, although alternative definitions of conjunction and disjunction can be defined.

New definitions

When a new rule is entered into the knowledge base, it is compiled into the code used to evaluate a query involving the new rule. For example, if we enter the new rule

TALL_PERSON(*x*) ← PERSON(*x*,*y*,_) & TALL(*y*)

this is compiled to the code

PROJ((1); JOIN((2,1)); PERSON(x,.,.), TALL

This code is almost identical to that used above in answering the query 'who is a tall person?'. The only difference is the inclusion of the variable x which occurs on the left-hand side of the rule. The inclusion of this variable is necessary to allow for selection when using the above rule. For example, the query

'Is Mary a tall person?'

is translated to

DOES (TALL_PERSON(=MARY))

which is compiled to

MAX–χ (TALL_PERSON(MARY))

and evaluated to give 0.5

which represents the truth that Mary is a tall person. The evaluation is as for the example above with the difference that:

(1) Only the (MARY ---) tuple of the PERSONS relation is used; this corresponds to the selection.
(2) The DOES query returns a χ value rather that a relation which the WHICH query returned. This χ value is determined as the maximum χ value associated with the relation returned by the WHICH query.

In the evaluation of MAX_χ the argument TALL_PERSON(MARY) is first evaluated. This is done by searching the dictionary for a definition of TALL_PERSON() and evaluating that. So that

PROJ((1); JOIN(((2,1)); PERSON(MARY,.,.), TALL))

is evaluated. PERSON(MARY,.,.) is evaluated to the tuple of PERSONS containing MARY and the remainder of the evaluation is as before but only with the one tuple of the PERSONS relation.

We can distinguish between fuzzy and non-fuzzy queries of the WHICH and DOES type as follows

Which query:

For non-fuzzy case: WHICH(x LIKES(x,y)) is a λ calculas statement $\lambda x \exists y$ LIKES(x,y) and a denotation operator δ returns $[x]$ such that $\exists y$ LIKES(x,y) is TRUE w.r.t. the data-base.

For the fuzzy case: If R = WHICH(x LIKES(x,y)) then R = PROJ((1); LIKES) where $\chi_{\mathrm{PROJ(1)LIKES}}(x) = \bigvee_{y} \chi_{\mathrm{LIKES}}(x,y)$

Does query:

For the non-fuzzy case: DOES($R(x,y,z)$) attempts to prove that $\exists x \exists y \exists z\ R(x,y,z)$ w.r.t. the data-base. Once an example is found proving this the query is given the answer YES.

For the fuzzy case: DOES($R(x,y,z)$) = MAX_ χ(WHICH((x,y,z),$R(x,y,z$))) and the query cannot be answered until all possible satisfying tuples have been found, to ensure that the highest χ value possible is returned. Thus for the fuzzy case it is necessary to determine all solutions to a query, even for the DOES type. In practice, of course, no more solutions for the DOES query need to be found if a χ value of 1 is returned from some tuple.

Another example

Knowledge base:

Base relations LIKES, PERSONS as above and

Tall	Ht	χ
I-type	5 – 5	0
	6 – 0	1.0
	7 – 0	1.0

FRIENDS(x,y) ← LIKES(x,y) & LIKES(y,x)
JOIN(((1,2),(2,1), LIKES(x,y), LIKES(y,x))

POSS_ATHLETE(x) ← PERSONS(x,y,_) & TALL(y)
PROJ((1), JOIN(((2,1)), PERSONS(x,.,.), TALL)

HAS_GOOD_FRIENDS(x) ← FRIENDS(x,y) & POSS_ATHLETE(y)
PROJ((1), JOIN(((2,1)), FRIENDS(x,.), POSS_ATHLETE)

Queries:

WHICH(x FRIENDS(x, =JON))

is compiled to

PROJ((1), FRIENDS(., JON))

which is evaluated to give

Sol	Name	χ
	Mary	0.7

WHICH((*x*,*y*) HAS_GOOD_FRIENDS(*x*) & PERSONS(*x*, _,*y*))

is compiled to

PROJ((1,3), JOIN(((1,1)), HAS_GOOD_FRIENDS, PERSONS)

and evaluated to give

Sol	Name	Wt	χ
	Jon	12 – 0	0.7
	Mary	11 – 5	0.6

Other aspects of FRIL

No attempt is made here to give a formal definition of the FRIL language or even to explain any other than its simplest aspects. The language does allow set theoretic relations to be defined as illustrated by the following example.

Knowledge base:

Has_ Ht	Name	Ht
	John	60
	Bill	30
	Mary	65
	Irene	32

LESS(x,y) = $\{(x,y) \mid x < y\}$

TALLER_THAN_SOMEONE(*x*) ← HAS_HT(*x*,*h*1) & HAS_HT(*y*,*h*2) & LESS(*h*2,*h*1)

PERSON(*x*) ← HAS_HT(*x*,_)

SHORTEST ← PERSON – TALLER_THAN_SOMEONE

SHORTEST_PERSON ← SHORTEST

Query:

'How tall is the shortest person?'

is asked by entering the query

WHICH(x SHORTEST(y) & HAS_HT(y,x))

and the solution

Sol	Ht
	30

is returned.

In this example, the LESS relation is defined in set theoretic terms, and the system is provided with an algorithm which for any given tuple (x,y) returns a χ-value of 1 if $x < y$, and 0 otherwise. The example also shows the use of the difference operator for relations. Many more constructs are available in FRIL including the intersection, union and complementation of relations, constructs for modifying the knowledge base through both the addition and deletion of facts and rules and the ability to define relations recursively. Different fuzzy logics can be used provided they are truth functional. For example, the definition of conjunction and disjunction can be changed to

$$v(A \text{ and } B) = v(A) \cdot v(B)$$

$$v(A \text{ or } B) = v(A) + v(B) - v(A) \cdot v(B)$$

or in fact to any T-norm logic discussed in the next section. By introducing a suitable logic, optimal path analysis problems can be solved in which the χ value is interpreted as the minimum distance between two nodes of a graph. Furthermore, FRIL is a module within a Prolog environment and messages in the form of queries can be passed to the FRIL module. The solution to the query in the form of a list of tuples representing the solution relation is then passed back from FRIL to the Prolog system. The next version of FRIL, suitable for concurrent processing and the exploitation of new computer architectures, will contain the present version of FRIL and Prolog as special cases and will be a true Logic Programming language.

Uncertainty logics

Fuzzy and probabalistic uncertainty logics have been discussed by Gaines[10] who put forward the idea that a basic uncertainty logic can be defined in terms of a valuation on a lattice of propositions. For any particular knowledge base, the uncertainty measure of a particular proposition may not be able to be given a point value.

Consider, for example, that a group of technologists vote on the suitability of a particular design from the point of view of it satisfying some specific specification. They must vote on whether or not the following proposition is true.

Design D1 is suitable

It is assumed that each member will vote true or false even though there will be doubt on its truthness since no design is wholly satisfactory. Each member will have a threshold level of satisfaction which is used to distinguish between true and false. Suppose that ⅔ of them vote true while the other ⅓ of them vote false. We can say that

$$v(\mathrm{P1}) = 2/3$$

Suppose further that the same group is asked about the design D2

P2: Design D2 is suitable

and that ⅓ of the group vote true and ⅔ vote false. Then

$$v(\mathrm{P2}) = 1/3$$

No record of their individual votings is taken. The only information recorded in each case is the proportion of people who voted true. What is the truth of the proposition

P3: Design D1 and design D2 are both suitable

and the proposition

P4: Design D1 or design D2 will be suitable.

We cannot give a point value to v(P1 and P2) or v(P1 or P2) but can give the following interval constraints for their values

$$0 \leqslant v(\mathrm{P1\ and\ P2}) \leqslant 1/3$$
$$2/3 \leqslant v(\mathrm{P1\ or\ P2}) \leqslant 1$$

In order to reduce these intervals, further information must be supplied or assumptions about the voting behaviour of the group made. It might be assumed, for example, that each individual retains the same threshold level from one decision to the next. With this assumption

$$v(\mathrm{P1\ and\ P2}) = \min(v(\mathrm{P1}), v(\mathrm{P2})) = 1/3$$
$$v(\mathrm{P1\ or\ P2}) = \max(v(\mathrm{P1}), v(\mathrm{P2})) = 2/3$$

Other assumptions will lead to alternative definitions.

Theory of dispositions

Zadeh introduced the concept of a disposition for a proposition which is almost always true. This motivates the use of fuzzy quantifiers such as MOST FEW etc. and requires the treatment of fuzzy probabilities. In addition, lack of information when inferring use of such propositions may require that fuzzy intervals instead of point values are associated with the conclusion of the inference. Statements of this type often occur as background knowledge in information systems. Consider the example

Most long term or violent prisoners return to prison within two years of being released.

Many of the prisoners in Jail X are long term or violent prisoners.

---

y% of prisoners in Jail X will return to prison within two years of being released.

What is *y* equal to? Once again a point value cannot be given but *y* is a possibility distribution on the interval [0,1], i.e., a fuzzy set on probability space. This problem can be rewritten as

Most (*A* or *B*)'s are *C*'s
Many *X*'s are (*A* or *B*)'s

---

y% of *X*'s are *C*'s

which can be further transformed to

a D's are *C*'s
b X's are *D*'s

---

g X's are *C*'s

and hence to

C :- *D* (*a*)
D (*b*)

C (*g*)

where a = Prob(C|*D*), b = Prob(D) and g = Prob(C). It is easily calculated that

$$ab \leq g \leq ab + 1 - b$$

Furthermore if

Prob(C) ∈ [Lower_Prob(C), Upper_Prob(C)]

then

$$\text{Prob(not C)} \in [1 - \text{Upper\_Prob(C)},\ 1 - \text{Lower\_Prob(C)}].$$

In the case when a and b are fuzzy probabilities, the interval is determined using

$$\text{Lower\_Prob}(C) = \mathbf{a} \otimes \mathbf{b}$$
$$\text{Upper\_Prob}(C) = (\mathbf{a} \otimes \mathbf{b}) \oplus (1 \ominus \mathbf{b})$$

That is fuzzy arithmetic is used to evaluate the appropriate expressions.

A theory of uncertainty

Consider the problem

A (a)
B (b)

$$\text{MAX}(0, a+b-1) \leqslant \text{Prob}(A \text{ and } B) \leqslant \text{MIN}(a,b)$$
$$\text{MAX}(a,b) \leqslant \text{Prob}(A \text{ or } B) \leqslant \text{MIN}(1, a+b)$$

This can be seen by considering Figure 1 which has four containers connected together and holding some liquid. The top left container represents the proposition that both A and B are true. The two containers on the left represent that B is true. A quantity '1' of liquid is allowed to flow in the containers but is further constrained according to the given propositions. Thus the top two containers must have 'a' amount of liquid and only that amount while the left two must contain 'b'. No other constraints are imposed so that the liquid can still have some movement. This movement gives rise to the intervals above for Prob(A and B) and Prob(A or B).

Furthermore we can deduce that

$$\text{Prob(Not } A) = 1 - a$$
$$\text{Prob (Not } B) = 1 - b$$
$$\text{Prob}(A) = \text{Prob}(A \text{ and } B) + \text{Prob}(A \text{ and Not } B)$$
$$\text{Prob}(B) = \text{Prob}(A \text{ and } B) + \text{Prob(Not } A \text{ and } B)$$

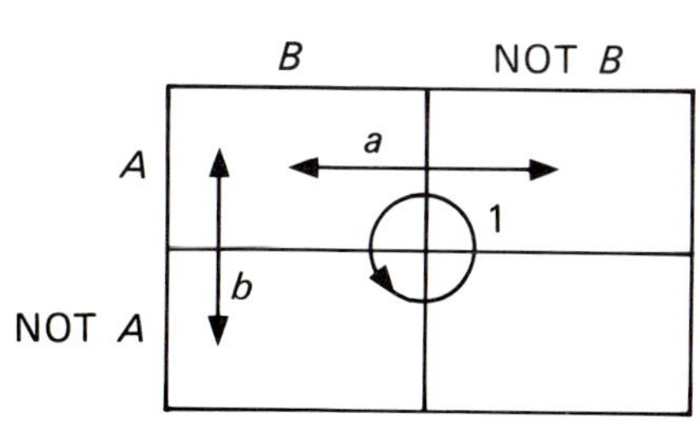

Figure 1 Theory of uncertainty using four containers holding some liquid.

If we make further assumptions then further restrictions for Prob(A and B) and Prob(A or B) can be given.

Further restrictions for conjunction:

IF A and B are mutually exclusive ($a + B \leq 1$ for this to be true)
THEN Prob(A and B) = 0

IF A and B are statistically independent
THEN Prob(A and B) = Prob(A).Prob(B)

IF one of A and B strictly implies the other
i.e. Prob(A and Not B) = 0 or Prob(Not A and B) = 0
THEN Prob(A and B) = MIN(Prob(A),Prob(B))

Further restrictions for disjunction:

IF one of A or B must occur
i.e. Prob(Not A and Not B) = 0
($a + b \geq 1$ for this to be possible)
THEN Prob(A or B) = 1

IF A and B are statistically independent
THEN Prob(A or B) = Prob(A) + Prob(B) − Prob(A and B)

IF one of A and B strictly implies the other
THEN Prob(A or B) = MAX(Prob(A),Prob(B)).

A general theory of uncertainty can be developed along these lines and special syllogisms stated with the conclusion given a probability belonging to an interval. This interval is easily calculated for any particular inference and will not be given in detail here. If we had made the assumption that the propositions must have truth values of 0 or 1 then the ordinary propositional calculus would result. If we had made the choice that the law of excluded middle and also, therefore, the law of contradiction must hold then the Rescher's probability logic would have resulted. Note that with the strict implication assumption given above a fuzzy logic corresponding to Lukasiewicz's multi-valued logic resulted. We might also require the assumption of truth functionality. Let us further suppose that T is a mapping

$$T: [0,1] \times [0,1] \rightarrow [0,1]$$

which satisfies the axioms

(1) $T(a,1) = a$
(2) $T(a,b) = T(b,a)$
(3) $T(a,b) \geq T(c,d)$ if $a \geq c$ and $b \geq d$
(4) $T(a,T(b,c)) = T(T(a,b),c)$

then T is called a *t*_norm and generalizes the 'AND' corresponding to conjunction.

Examples of instances of T are

(1) $T(a,b) = \mathrm{MIN}(a,b)$
(2) $T(a,b) = a,b$
(3) $T(a,b) = 1-\mathrm{MIN}(1((1-a)^p + (1-b)^p)^{1/p});\ p \geqslant 1$

A dual norm called the *t*_co-norm S exists which generalizes disjunction. For any *t*_norm T there exists a dual norm S such that

$$S(a,b) = 1 - T((1-a), (1-b))$$

The mapping S: $[0,1] \times [0,1] \rightarrow [0,1]$ satisfies the same axioms as for T except that (1) is replaced by (1′) where (1′) $S(a,0) = a$

Networks

Consider the following logic inference

C:–	(A or B)	;	($\alpha = .7$)
E:–	(D and C)	;	($\beta = .9$)
A		;	($a = .7$)
B		;	($b = .8$)
D		;	($d = .6$)

C
E

and the corresponding inference network is shown in Figure 2.

We can calculate the uncertainty intervals associated with the various probabilities as shown in Figure 2. Thus we obtain that

$$0.56 \leqslant \mathrm{Prob}(C) \leqslant 0.76$$

and

$$0.144 \leqslant \mathrm{Prob}(E) \leqslant 0.984$$

so that very little can be concluded about Proposition E. We might, of course, have made invalid or unjustifiable assumptions in order to have arrived at a point value for these probabilities and thus made an error of judgement.

In making deductions using background knowledge, long chain arguments involve large uncertainty intervals. Human capability is limited to short chain arguments, minimizing the danger of deducing what is really uncertain. When using the computer for such inferences, it is not limited in this way and one must avoid the danger of error by using interval analysis as illustrated above.

A serious difficulty and one we have not yet considered is how can one combine evidences from different sources to arrive at a final conclusion. Consider, for example, the background statement

Most Americans of age 40 live at least another 10 years.
Few lung cancer patients live more than a few years after diagnosis.

with the specific statements

John is a 30 year old American.
John has been diagnosed as having lung cancer.

What conclusion can we make about the probability that John will live another 10 years?

This difficulty is further illustrated with a modified form of Prolog which we will call Probabalistic Prolog. The following example illustrates the point

```
heavy(x) :- person(x,h), tall(h), not(thin(x)) ; (γ)
person(jon,6)
tall(6)
not_thin(x) :- shoe_size(x,y), large (y) ; (β)
shoe_size(jon,11)
large(11)
thin(x) :- not(not_thin(x))
?_ heavy(jon)
yes[γβ,γβ+1−β]
?_not(heavy(jon))
yes [β(1−γ),1−γβ]
```

In this case, a query is evaluated in the normal way but a probability for the proof path is then calculated. This, in general, will be an interval and found by methods outlined above.

In this case the query

?—heavy(jon)

results in a probability interval $[\gamma\beta, \gamma\beta+1-\beta]$ and the query

?—not heavy(jon)

results in the probability interval $[\beta(1-\gamma), 1-\gamma\beta]$.

Note that these upper and lower probabilities satisfy the relationship of upper and lower probabilities of the proposition and its negation given previously.

In this example, there is only one way of proving the theorem in each

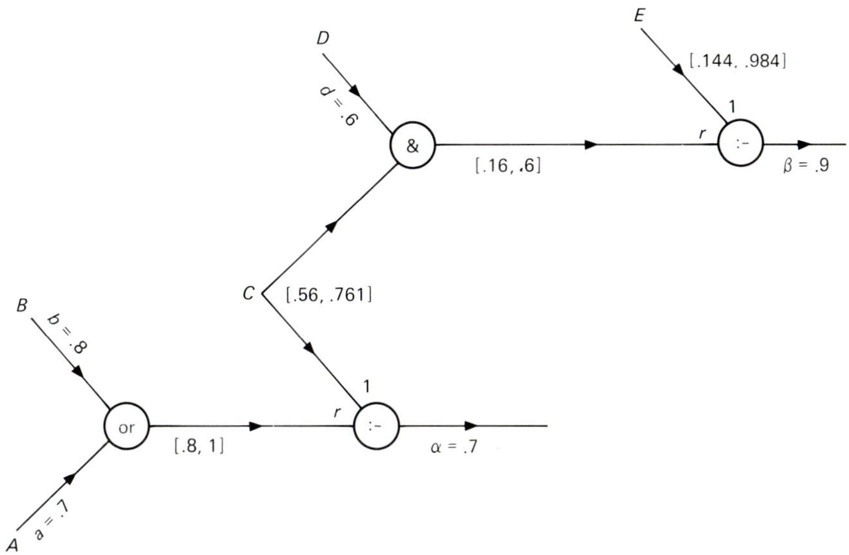

Figure 2 Interference network.

case. In general, there will be more than one proof path for answering a query and a probability interval can be associated with each of the paths resulting in the query being answered. The problem remains of how to combine these intervals to arrive at the final conclusion of giving one probability or probability interval to the solution of the query.

References

1. Russell, B., 'Vagueness', *Austr. J. Psychol. Phil., 1*, 84–92 (1923).
2. Bellman, R.E., 'Mathematics, Systems and Society', *FEK Rep.*, Stockholm (1971).
3. Baldwin, J.F., and Guild, N.C.F., 'The Resolution of two Paradoxes by Approximate Reasoning using a Fuzzy Logic', *Synthese, 44*, 397–420 (1980).
4. Baldwin, J.F., 'A new Approach to Approximate Reasoning using a Fuzzy Logic', *Fuzzy Sets Systems, 2*, 309–325 (1979).
5. Zadeh, L.A., 'Fuzzy Sets', *Information Control, 8*, 338–353 (1965).
6. Baldwin, J.F., 'A Fuzzy Relational Inference Language for Expert Systems', Proc. Int. Symp. on Multi-valued Logic, Japan (1983).
7. Zadeh, L.A., 'PRUF—A Meaning Representation Language for Natural Languages', in Mamdani, E., and Gaines, B. (Editors), *Fuzzy Reasoning and It's Applications*, Academic Press, New York (1981).
8. Ullman, J.D., *Principles of Data Base Systems*, Pitman, New York (1980).
9. Kowalski, R., *Logic for Problem Solving*, North Holland, Amsterdam (1979).
10. Gaines, B.R., 'Fuzzy and Probability Uncertain Logics', *Information Control, 38* (2), 154–169 (1978).

Professor Henri Nussbaumer

Henri Nussbaumer is Computer Science Professor at l'Ecole Polytechnic Fédérale de Lausanne. He graduated as an engineer from l'Ecole Centrale de Paris in 1954, and later received his Doctorat d'Etat from Nice University. He was employed at the IBM France Laboratory from 1957 to 1981 and appointed IBM Fellow in 1975. He joined l'Ecole Polytechnique Federal de Lausanne as Computer Science Professor in 1981, and in 1983 was elected Fellow of the Institute of Electrical and Electronic Engineers.

His main fields of activity have been computer communications, digital signal processing, computational complexity, and fast algorithms. He is author or co-author of some 40 scientific papers and 56 patents on signal processing, data communications, and electronic line switching. He has also published a book on fast algorithms.

Digital signal processing and computational complexity

The efficiency of digital signal processing algorithms has become an issue of growing importance, owing to the rapid development of real time applications. There are a number of ways to characterize the efficiency of an algorithm. From the standpoint of the final user, an algorithm is efficient when it yields a cost-effective solution with existing hardware, and in a sense, this definition of the algorithm efficiency is the only one that counts in practice. Unfortunately, while such a formulation provides the final yardstick for evaluating the efficiency of an algorithm, it does encompass too many factors to provide a framework for developing new algorithms. One is, therefore, led to consider separately the complexity of the algorithms and the efficiency of their implementation in a processor. By successive iterations, it is then possible to design better signal processors and time efficient algorithms.

The present discussion is restricted to DFT (discrete Fourier transform) and convolution algorithms, along with some related algorithms. First we consider the algebraic complexity of such algorithms, that is, the number of moves or arithmetic operations required to evaluate DFTs and convolutions. We will then discuss the arithmetic techniques that may be used to further reduce the complexity. Finally, we will briefly address the issues related to the implementation of digital filter banks.

DFT algorithms

The DFT plays a major role in many digital signal processing applications. The motivation for developing efficient DFT algorithms is related to the fact that the direct computation of an N-point DFT is of the order of N^2 (denoted $0\,(N^2)$) operations, and, therefore, becomes rapidly excessive for large dimensions. The most important event in the development of efficient DFT algorithms has been the introduction of the fast Fourier transform (FFT)[1] which reduces the DFT computational complexity to $0\,(N\log N)$. This algorithm is based on a divide and conquer strategy, and may be described in polynomial algebra as follows..

Let us consider the DFT $\bar{X}_k$ of an N-point sequence x_m, with

$$\bar{X}_k = \sum_{m=0}^{N-1} x_m W^{mk}, \quad k = 0,\ldots,N-1$$

$$W = e^{-j2\pi/N}, \quad j = \sqrt{-1} \tag{1}$$

In polynomial algebra, this DFT may be represented by the two following equations

$$X(z) = \sum_{m=0}^{N-1} x_m z^m \text{ Modulo } (z^N-1) \tag{2}$$

$$\bar{X}_k = X(z) \text{ Modulo } (z-W^k) \tag{3}$$

where equation (3) means that z is replaced by W^k. It may be verified easily by replacing z by W^k in equation (2) that one obtains equation (1), and therefore that equations (2) and (3) are a valid representation of equation (1). Since the N polynomials $z-W^k$, for k=0,..., $N-1$ are the N factors of z^N-1, the DFT computation may be organized by first reducing $X(z)$ modulo one of the factors $P_i(z)$ of z^N-1, and then reducing the resulting polynomial modulo $z-W^k$, with k restricted to the set of integers which correspond to the roots W^k of $P_i(z)$:

$$\bar{X}_k = \Big[[[X(z) \text{ Modulo } (z^N-1)] \text{ Modulo } P_i(z)]\Big]$$

$$\text{Modulo } (z-W^k),$$

$$W^k \in \text{set of roots of } P_i(z) \tag{4}$$

With this method, which is the basis for most fast DFT algorithms, the computation of $\bar{X}_k$ is divided among the factors $P_i(z)$ of z^N-1. In practice, the various algorithms differ by the factorization of z^N-1, which in turn depends on the field of coefficients. In the case of the conventional FFT algorithm, z^N-1 is factorized into irreducible polynomials having factors that are rational numbers. These polynomials are called cyclotomic polynomials, and it can be shown that the number of distinct cyclotomic polynomials which are the factors of z^N-1 is equal to the number of divisors of N, including 1 and N. This number is maximum when $N = 2^t$. Thus, one obtains the best factorization and therefore the best computing efficiency when $N = 2^t$. In this case, the decomposition may be done by first reducing z^N-1 into the two factors $z^{N/2}+1$ and $z^{N/2}-1$, with

$$X_1(z) = \sum_{m=0}^{N/2-1} (x_m - x_{m+N/2})\, z^m$$

$$\equiv X(z) \text{ Modulo } (z^{N/2}+1) \tag{5}$$

$$\bar{X}_{2u+1} \equiv X_1(z) \text{ Modulo } (z-W^{2u+1}),$$
$$u = 0,\ldots,N/2-1 \tag{6}$$

and

$$X_2(z) = \sum_{m=0}^{N/2-1} (x_m + x_{m+N/2})z^m$$
$$\equiv X(z) \text{ Modulo } (z^{N/2}-1) \tag{7}$$
$$\bar{X}_{2u} \equiv X_2(z) \text{ Modulo } (z-W^{2u}),$$
$$u = 0,\ldots, N/2-1 \tag{8}$$

Equations (7) and (8) represent a DFT of size $N/2$ on which the algorithm may be repeated. Equations (5) and (6) represent an odd DFT of size $N/2$, on which the algorithm cannot be repeated without some modification, because $z^{N/2}+1$ is an irreducible polynomial for coefficients defined in the field of rational numbers. However, the decomposition of equations (5) and (6) may be pursued if the sequence $x_m - x_{m+N/2}$ is multiplied by W^m, which is equivalent to replacing z by Wz in equations (5) and (6). With this modification, equations (5) and (6) become

$$X_1(z) = \sum_{m=0}^{N/2-1} (x_m - x_{m+N/2})W^m z^m$$
$$\equiv X(Wz) \text{ Modulo } (z^{N/2}-1) \tag{9}$$
$$\bar{X}_{2u+1} = X_1(z) \text{ Modulo } (z-W^{2u}), \tag{10}$$

which represents a DFT of size $N/2$. Hence an N-point DFT is converted into two ($N/2$)-point DFTs at the cost of $N/2$ complex multiplications by W^m and N additions. Therefore, with complete decomposition, the DFT is computed with $N\log_2 N$ additions and $(N/2)\log_2 N$ multiplications. One can see that the set of additions and subtractions in each stage of the FFT algorithm may be viewed as resulting from reductions modulo (z^u-1), (z^u+1), while the multiplications by the twiddle factors may be considered as a way of converting a field defined modulo $z^{N/2}+1$ into a ring defined modulo $z^{N/2}-1$. The FFT algorithm has an arithmetic complexity which is $0(N\log N)$, with a number of moves which is also $0(N\log N)$. Moreover, when the same decomposition is used at each stage, as for instance with a radix-2 or radix-4 algorithm, the structure of the algorithm is simple and regular, with a minimum of ancillary operations. These desirable features, plus the availability of carefully optimized FFT programs[2] are the reasons why the various versions[3–5] of the FFT are the most widely used fast DFT algorithms.

When the FFT algorithms run in a processor where the multiplication time is significantly longer than the addition time, it is possible to implement the FFT with a complex multiplication method which requires three real multiplications and three real additions instead of four real multiplications and two real additions.[4] Another possibility to reduce the computation load related to complex multiplications consists of modifying the FFT algorithm in order to replace the complex twiddle factors by real twiddle factors. This may be done by the Bruun algorithm[6] which is discussed here in a modified version, in order to fit the general frame of this presentation.

In the conventional FFT decomposition, one has to introduce the multiplication by the twiddle factors W^m in equation (5) because the polynomial $z^{N/2}+1$ is irreducible with coefficients in the field of rational numbers. However, if we now work with irrational numbers, $z^{N/2}+1$ may be factorized, with

$$z^{N/2}+1 \quad = \quad (z^{N/4}+\sqrt{2}z^{N/8}+1)\,(z^{N/4}-\sqrt{2}\,z^{N/8}+1) \tag{11}$$

With this technique, equations (5) and (6) are computed with

$$X_3(z) \quad \equiv \quad X_1(z) \text{ Modulo } (z^{N/4}+\sqrt{2}\,z^{N/8}+1) \tag{12}$$

$$\bar{X}_{2u+1} \quad \equiv \quad X_3(z) \text{ Modulo } (z-W^{2u+1})\,,$$

$$W^{2u+1} \in \text{set of roots of } (z^{N/4}+\sqrt{2}\,z^{N/8}+1), \tag{13}$$

with similar relations for the decomposition modulo $z^{N/4} - \sqrt{2}\,z^{N/8}+1$. With this technique, the multiplications by complex twiddle factors are replaced with the multiplications by the real factor $\sqrt{2}$ resulting from the reduction in equation (12). The real decomposition may be pursued in a similar way by noting that

$$z^{4q} + a\,z^{2q} + 1 = (z^{2q} + \sqrt{2-a}\,z^q + 1)$$
$$(z^{2q} - \sqrt{2-a}\,z^q + 1) \tag{14}$$

There are many variations of the real factor FFT algorithm.[7,8] These algorithms do not reduce significantly the number of arithmetic operations over a conventional radix-4 FFT algorithm implemented with a complex multiplication method using three real multiplications and three real additions. However, they may be more efficient than conventional algorithms for computing the DFT of real sequences in a microcomputer with a limited number of registers.

The FFT decomposition process would convert a one-dimensional DFT into a multidimensional DFT, were it not for the twiddle factors that creep at each stage. When the DFT size N is the product of d relatively prime factors $N_1, N_2, \ldots N_d$, a true decomposition into a d-dimensional DFT is possible, and this provides a new set of fast DFT algorithms that differ drastically from the FFT. This one-dimensional to multidimensional mapping is done by simple index transformations[9,10] such as

$$m = N_1 m_2 + N_2 m_1 \text{ Modulo } N ,$$

$$m_1 = 0, \ldots, N_1-1 , \quad m_2 = 0, \ldots, N_2-1 \qquad (15)$$

for $N = N_1 N_2$, $(N_1 N_2) = 1$, and with a similar relation for k. With this mapping, and for instance, $d = 2$, the one-dimensional DFT (1) is converted into a two-dimensional DFT of size $N_1 \times N_2$

$$\bar{X}_{k1,k2} = \sum_{m_1=0}^{N_1-1} \sum_{m_2=0}^{N_2-1} x_{m_1,m_2} W_1^{m_1 k_1} W_2^{m_2 k_2} ,$$

$$W_1 = e^{-j2\pi/N_1} , \quad W_2 = e^{-j2\pi/N_2} ,$$

$$k_1 = 0, \ldots, N_1-1, \quad k_2 = 0, \ldots, N_2-1 \qquad (16)$$

At this point, the multidimensional DFT may be computed either by a conventional multidimensional DFT technique,[9–11] or by a new algorithm introduced by Winograd.[12] In the first case, now called the prime factor technique, the calculations are carried out separately along the various dimensions, for instance, along the lines first, and then along the columns for the two-dimensional DFT.[16] Thus, if Mu, Mu_1, Mu_2 and Ad, Ad_1, Ad_2 are the number of complex multiplies and adds required to compute DFTs of size N,N_1 and N_2 respectively, we have

$$Mu = N_1 Mu_2 + N_2 Mu_1 \qquad (17)$$

$$Ad = N_1 Ad_2 + N_2 Ad_1 \qquad (18)$$

With the Winograd Fourier transform algorithm (WFTA), the multidimensional DFT is computed by a nesting technique[12] which, in the case of the two-dimensional DFT (16) amounts to calculating a N_1-point DFT in which scalars are replaced by vectors of N_2 points and in which each scalar multiplication is replaced by the computation of a N_2-point DFT. For a two-factor decomposition, this yields a number of operations given by

$$Mu = Mu_1 Mu_2 \qquad (19)$$

$$Ad = Mu_1 Ad_2 + N_2 Ad_1 \qquad (20)$$

In order to have an efficient algorithm with the prime factor and Winograd techniques, it is necessary to choose a DFT size N such that N has many relatively prime factors, so as to reduce the computation of a large DFT into that of many small DFTs. The small DFTs may in turn be computed by carefully optimized algorithms that generally use a technique based on the conversion of DFTs into real factor convolutions,[13] and which is very efficient for small DFTs (and relatively inefficient for large DFTs). With these methods, the Winograd algorithm requires two to four times less multiplications than the FFT algorithm, and slightly more additions. In the case of a 1,008-point DFT, for instance, the Winograd algorithm requires 3.52 real multiplications and 34.39 additions per output sample, as opposed to 7 real multiplications and 33.25 real additions for a real factor 1,024-point FFT algorithm.[4] The prime factor algorithm requires usually more multiplications but less additions than the Winograd algorithms. Thus, the Winograd and the prime factor algorithms decrease significantly the number of multiplications over the conventional FFT.

However, these algorithms are not yet in frequent use, owing partly to the fact that they have been introduced (or reintroduced) more recently than the FFT. A key issue in the application of the WFTA concerns its ability to be translated into computer programs that are more effective than conventional FFT programs. This matter is still debated, but it seems that the WFTA is not usually more time efficient than the FFT, because of its more complex structure, and because the savings in number of multiplications are offset by the increase in number of additions on computers where the speed of the two operations is about the same.[14,15] It has been shown, however, that the lower bound on the number of multiplications required to compute an N-point DFT is $0(N)$, and research is going on to develop a practical algorithm which would achieve this lower bound with a regular structure.

Multidimensional DFT algorithms

Multidimensional DFTs may be computed conventionally either by the Winograd algorithm or by a row–column method that is similar to the prime factor technique described above. In this last case, a DFT of size $N \times N$ is computed with N DFTs along the rows and N DFTs along the columns, with a total of $N^2 \log_2 N$ complex multiplications and $2N^2 \log_2 N$ complex additions, if the calculations are carried out with a radix-2 FFT algorithm.

When there are common factors in the various dimensions, which is generally the case, a more efficient technique may be used by resorting to polynomial transforms.[4,16] The polynomial transform technique may be described simply, in the case of a DFT of size $N \times N$, with $N = 2^t$, defined by

$$\bar{X}_{k1,k2} = \sum_{m_1=0}^{N-1} \sum_{m_2=0}^{N-1} x_{m_1,m_2} W^{m_1 k_1} W^{m_2 k_2} ,$$

$$W = e^{-j2\pi/N} , \quad k_1, k_2 = 0, \ldots, N-1 \tag{21}$$

This DFT may be represented in polynomial algebra, with a technique similar to that used in equations (2) and (3), with

$$X_{m_1}(z) \quad = \quad \sum_{m_2=0}^{N-1} x_{m_1,m_2} z^{m_2} \tag{22}$$

$$\bar{X}_{k_1}(z) \quad \equiv \quad \sum_{m_1=0}^{N-1} X_{m_1}(z)\, W^{m_1 k_1} \text{ Modulo } (z^N - 1) \tag{23}$$

$$\bar{X}_{k_1,k_2} \quad \equiv \quad \bar{X}_{k_1}(z) \text{ Modulo } (z - W^{k_2}) \tag{24}$$

The computation is then split into two parts which correspond to the factors $(z^{N/2}-1)$ and $(z^{N/2}+1)$ of z^N-1. The first part corresponds to a DFT of size $N\times(N/2)$ on which the algorithm is repeated. The second part is computed by introducing a permutation k_2k_1, using the fact that k_2 is odd, since all the roots of $z^{N/2}+1$ are given by W^{k_2}, for k_2 odd. With this permutation, equations (23) and (24) become

$$\bar{X}_{k_2k_1}(z) \quad = \quad \sum_{m_1=0}^{N-1} X_{m_1}(z)\, W^{m_1 k_2 k_1} \text{ Modulo } (z^{N/2}+1) \tag{25}$$

$$\bar{X}_{k_2k_1,k_2} \quad = \quad \bar{X}_{k_2k_1}(z) \text{ Modulo } (z - W^{k_2}) \; k_2 \text{ odd} \tag{26}$$

Equation (25) may be simplified by noting that equation (26) implies that $z = W^{k_2}$

$$\bar{X}_{k_2k_1}(z) \quad = \quad \sum_{m_1=0}^{N-1} X_{m_1}(z)\, z^{m_1 k_1} \text{ Modulo } (z^{N/2}+1) \tag{27}$$

Equation (27) is a transform which we have called a polynomial transform and which is computed without multiplications.

With this polynomial transform technique, about half the multiplications required by the conventional row–column FFT algorithm are eliminated, and some additions are also saved. A similar technique can be used to compute multidimensional discrete cosine transforms (DCT).[17] Moreover, the polynomial transform method may be combined with the Winograd algorithm to yield larger computational savings. In this last case, the number of multiplications can be very low, as exemplified by a DFT of size 1,008 × 1,008 which is computed with only 3,39 real multiplications per output sample. Polynomial transforms can be computed with a regular, FFT-type structure which can be implemented easily. These features have been reported to yield execution times two to four times shorter than with the conventional FFT,[18] owing partly to the fact that the polynomial transform approach eliminates matrix transpose operations.

Convolution algorithms

The convolution y_r of two N-point sequences x_m and h_n is given by

$$Y_r = \sum_{n=0}^{N-1} h_n x_{r-n} \tag{28}$$

where $r-n$ is defined modulo N when the convolution is cirçular. In polynomial notation, this convolution is expressed by

$$Y(z) = H(z)\,X(z) \tag{29}$$

with

$$X(z) = \sum_{m=0}^{N-1} x_m z^m \tag{30}$$

and with a similar relation for $H(z)$. If the convolution is circular, $Y(z)$ is defined modulo z^N-1. Most algebraic convolution algorithms may be viewed as derived from the Chinese remainder[4] which states that if $Y(z)$ is defined modulo a polynomial $P(z)$, it can be reconstructed from the d residues $Y_k(z)$ modulo the d relatively prime factors $P_k(z)$ of $P(z)$, with

$$Y(z) \equiv \sum_{k-1}^{d} Y_k(z)\,S_k(z) \text{ Modulo } P(z) \tag{31}$$

where

$$S_k(z) \equiv 1 \text{ Modulo } P_k(z)$$

$$S_k(z) \equiv 0 \text{ Modulo } P_i(z)\,, \quad i \neq k \tag{32}$$

When the convolution is circular, $Y(z)$ is defined modulo $P(z) = z^N-1$ which factors into the N polynomials $z-W^k$, with $k = 0, \ldots, N-1$. In this case, the computation by the Chinese remainder theorem is strictly identical to the conventional computation of a circular convolution by DFTs, since one must compute the DFTs $H(W^k)$ and $X(W^k)$, multiply $X(W^k)$ by $H(W^k)$ and reconstruct $Y(z)$ by equation (31), which equivalent to calculating an inverse DFT, since $S_k(z) = (z^N-1)/(zW^{-k}-1)N$. When the convolution is aperiodic, $Y(z)$ is not changed if it is defined modulo any polynomial of degree equal or greater than $2N-1$, since the degree of $Y(z)$ is equal to $2N-2$. Under this condition, one may chose $P(z) = z^{2N-1}-1$ and compute the aperiodic convolution by DFTs of size $2N-1$. This method, which is nothing else but the conventional FFT computation of aperiodic convolutions by the overlap-add algorithm,[4] is the most frequently used fast convolution algorithm.

In many instances, it is desirable to do away with the ancillary operations and computations on complex numbers required with the FFT. This may be done for aperiodic convolutions, by continuing to use the Chinese remainder theorem approach, but with a polynomial $P(z)$ which factors into $2N-1$ polynomials $z-a_k$ where the a_k are $2N-1$ distinct integers. This approach requires theoretically only $2N-1$ multiplications, but in practice, it is efficient only for small convolutions, because the complexity of the reductions and the Chinese reconstruction increases rapidly with the convolution size. It is, however, possible to devise efficient algorithms for convolutions of sizes of up to about $N = 400$, by using an index mapping and a nesting technique similar to the Winograd algorithm, which reduces the computation of a long convolution into that of many small convolutions, for which efficient algorithms exist.[19]

Multidimensional convolutions may be computed by multidimensional DFTs or by a nesting technique.[19] However, when there are common factors among the various dimensions, polynomial transforms provide an efficient way of mapping without multiplications the multidimensional convolutions into one-dimensional convolutions.[4,20] The resulting algorithms have a significantly lower operations count than does the conventional convolution algorithm with a regular FFT-type structure. These algorithms may be derived from the polynomial transform DFT algorithms, and it can be shown that the polynomial transforms share with the DFT the convolution property, but with scalar multiplications in the frequency domain replaced by polynomial multiplications.

Arithmetic algorithms

Up to now, we have restricted our discussion to algebraic algorithms. Different algorithms may also be devised by including the arithmetic operations performed on the signal.

A first class of algorithms may be obtained by constraining all the arithmetic operations to be performed modulo some integers. Among the many algorithms based on residue arithmetic, the most interesting seem to be those which are based on the convolution property of NTTs (Number Theoretic Transform).[21,22] These NTTs have the same structure as a DFT or a polynomial transform, but with the root W or z replaced by an integer, and with all operations performed modulo another integer. The best NTTs are those for which the multiplications by the powers of the root are simple, with simple operations also for the reductions. These conditions are met with the Mersenne and Fermat transforms which are defined by

$$\bar{X}_k = \sum_{m=0}^{N-1} x_m \, 2^{mk} \text{ Modulo } P \qquad (33)$$

where $p = 2^q + 1$, $q = 2^t$, $N = 2q$ for a Fermat transform and $p = 2^q - 1$, q prime, $N = q$ for a Mersenne transform. These NTTs are used exactly like DFTs to compute circular convolutions by calculating the NTTs of the input sequence, multiplying term by term the transform samples, and computing the inverse transform. If the multiplications by powers of 2 are much simpler than general multiplications, the NTTs may be considered as computed only with additions, and an N-point circular convolution is calculated with just N multiplications. The NTT circular convolution property is a consequence of the fact that a one-dimensional convolution on B-bit scalars is the two-dimensional convolution of binary vectors, with

$$x_m = \sum_{u=0}^{B-1} x_{m,u} 2^u, \quad x_{m,u} \in \{0,1\} \tag{34}$$

and a similar relation for representing h_n as a fonction of its bits $h_{n,v}$. Thus, the scalar convolution (28) may be expressed as a two-dimensional convolution with

$$Y_r = \sum_{b=0}^{2B-1} \left(\sum_{n=0}^{N-1} \sum_{v=0}^{B-1} h_{n,v} x_{r-n,b-v} \right) 2^b \tag{35}$$

This shows that the NTTs are particular cases of polynomial transforms where the B-bits words are viewed as polynomials.

The NTTs have not yet been widely used because most existing computers do not process efficiently powers of two shifts and modular arithmetic.

Signal processing at the bit level may be done with an entirely different class of algorithms called distributed arithmetic[23,24] which do away with the block processing and modular arithmetic used with NTTs. These algorithms are based on a rearrangement of the order of the summations and on the precomputation of the partial products in equation (35), which treats a convolution as a generalized multiplication and trades speed for memory. In practice, this technique allows a two to four speed improvement over direct computation, but has the disadvantage of requiring bit manipulations that are not easily implemented with most existing processors.

Banks of filters

Uniform digital bandpass filter banks are used extensively in many signal processing applications. These filter banks may be viewed as a set of bandpass filters with identical bandwidth and with centre frequencies uniformly spaced in the frequency domain. The complexity of such digital filter banks may be significantly reduced by using conventional fast convolution techniques. However, in this particular case, it is possible to take advantage of the regular filter bank structure to reduce the computation to that of a simple polyphase network, plus a DFT and some auxiliary operations.[25,26] Because the filters in the polyphase network are much simpler than the original bandpass filters, this approach reduces drastically the computational complexity. In essence, this polyphase technique, which has many variations, is based on the fact that the N bandpass filters with z-transform $G_k(z)$, are derived by frequency translation from a prototype filter $H(z)$, with

$$G_k(z) = H(W^{-k}z)\ , \quad W = e^{-j2\pi/N} \tag{36}$$

These filters may be split into N polyphase filters $H_n(z^N)$

$$G_k(z) = \sum_{n=0}^{N-1} z^{-n}W^{nk}H_n(z^N) \tag{37}$$

Thus, the N original L-tap bandpass filters are converted into N new (L/N)-tap polyphase filters and a DFT. This technique is, therefore, very efficient, and has successfully been applied to transmultiplexers for conversion between time-division multiplex systems and frequency division multiplex equipments.

In some applications, such as speech compression, the bank of N filters must be such that a signal which is split into N sub-bands subsampled by $1/N$ can be perfectly reconstructed from the N sub-bands. When conventional filter banks are used, this perfect reconstruction condition can be approximately met only with very high order filters, since the bandpass filters either overlap, which cause aliasing, or they do not overlap, and the signal is not perfectly reconstructed, because of the interchannel gaps. A very efficient solution to this problem using quadrature mirror filters (QMF) has been proposed.[27] The QMF technique results in a perfect signal reconstruction, regardless of the order of the filter, thus allowing the use of much shorter filters while improving performances. Moreover, QMF filters can be implemented efficiently by using a radix-2 tree decomposition technique.[27,28]

Concluding remarks

We have limited our discussion to a few selected convolution and DFT algorithms which have been developed either as *ad hoc* solutions to very specific problems or as part of a mathematical complexity theory based on conceptual models of complexity. From the standpoint of the end user, the ultimate test of efficiency of an algorithm is the execution time, or some other measure, such as memory occupancy, given the hardware and software environment. Considerable success has been achieved in devising algorithms that are efficient from the standpoint of the number of operations. However, new algorithms now need to be developed which are better adapted to existing hardware, or some new architectures devised that exploit more efficienctly existing signal processing algorithms. Work is progressing in both directions.[14,29,30] It seems, in particular, that systolic arrays may provide a very good implementation for some arithmetic-efficient algorithms based on NTTs or polynomial transforms.

References

1. Cooley, J.W., and Tukey, J.W., 'An algorithm for the machine calculation of complex Fourier series', *Math. Comput., 19*, 297–301 (1965).
2. IEEE, *Programs for Digital Signal Processing*, IEEE Press Book No. 0-87942-127-4, (1979).
3. Singleton, R., An algorithm for computing the mixed radix fast Fourier transform', *IEEE Trans. Audio Electroacoust., AU–17*, 93–103 (1969).
4. Nussbaumer, H.J., *Fast Fourier Transform and Convolution Algorithms*, 2nd Edn, Springer-Verlag, Berlin, Heidelberg, New York (1982).
5. Bergland, G.D., 'A radix eight fast Fourier transform subroutine for real valued series', *IEEE Trans. Audio Electroacoust., AU–17*, 138–144 (1969).
6. Bruun, G., 'Z. transform DFT filters and FFTs', *IEEE Trans. Acoust., Speech, Signal Processing, ASSP–26*, 56–63 (1978).
7. Rader, C.M. and Brenner, N.M., 'A new principle for fast Fourier transformation', *IEEE Trans. Acoust., Speech, Signal Processing, ASSP–24*, 264–265 (1976).
8. Cho, K.M. and Temes, G.C., 'Real factor FFT algorithms', *Proc. IEEE ICASSP*, 634–637 (1978).
9. Good, I.J., 'The interaction algorithm and practical Fourier analysis', *J. R. Stat. Soc., B20*, 361–372 (1958).
10. Burrus, C.S., 'Index mappings for multidimensional formulation of the DFT and convolution', *IEEE Trans. Acoust., Speech, Signal Processing, ASSP–25*, 239–242 (1977).
11. Kolba, D.P. and Parks, T.N., 'A prime factor FFT algorithm using high speed convolution', *IEEE Trans. Acoust., Speech, Signal Processing, ASSP–25*, 281–294 (1977).
12. Winograd, S., 'On computing the discrete Fourier transform', *Proc. Nat. Acad. Sci. USA, 73*, 1005–1006 (1976).
13. Rader, C.M., 'Discrete Fourier transforms when the number of data samples is prime', *IEEE Proc., 56*, 1107–1108 (1968).
14. Morris, L.R., 'A comparative study of time efficient FFT and WFTA programs for general purpose computers', *IEEE Trans. Acoust., Speech, Signal Processing, ASSP–26*, 141–150 (1978).
15. Nawab, H. and McClellan, J.H., 'Bounds on the minimum number of data transfers in WFTA and FFT programs', *IEEE Trans. Acoust. Speech, Signal Processing, ASSP–27*, 394–398 (1979).
16. Nussbaumer, H.J., and Quandalle, P., 'Fast computation of discrete Fourier transforms using polynomial transforms', *IEEE Trans. Acoust., Speech, Signal Processing, ASSP–27*, 169–181 (1979).

17. Nussbaumer, H.J., 'Fast polynomial transform computation of the 2-D DCT' *Proc. Int. Conf. Dig. Sign. Processing*, 276–283 (1981).
18. Kriz, T.A., 'Reduced data re-order complexity properties of polynomial transform 2D convolution and Fourier transform methods', *IBM J. Res. Develop., 26,* 708–714 (1982).
19. Agarwal, R.C. and Cooley, J.W., 'New algorithms for digital convolution', *IEEE Trans. Acoust., Speech, Signal Processing, ASSP–25,* 392–409 (1977).
20. Nussbaumer, H.J. and Quandalle, P., 'Computation of convolutions and discrete Fourier transforms by polynomial transforms', *IBM J. Res. Develop., 22,* 134–144 (1978).
21. Pollard, J.M., 'The fast Fourier transform in a finite field', *Math. Comput., 25* 365–374 (1971).
22. Agarwal, R.C. and Burrus, C.S., 'Number theoretic transforms to implement fast digital convolution', *Proc. IEEE, 63,* 550–560 (1975).
23. Croisier, A., Estban, D.J., Levilion, M.E., and Riso, V., 'Digital filter for PCM Encoded Signals', US Patent 3777130 (4 December 1973).
24. Burrus, C.S., 'Digital filter structures described by distributed arithmetic', *IEEE Trans., CAS–24,* 674–680 (1977).
25. Bellanger, M.G., and Daguet, J.L., 'TDM–FDM transmultiplexer: digital polyphase and FFT', *IEEE Trans. Commun., COM–22,* 1199–1205 (1974).
26. Scheuermann, M., and Glöckler, H., 'A comprehensive survey of digital transmultiplexing methods', *Proc. IEEE, 69,* 1419–1450 (1981).
27. Croisier, A., Esteban, D., and Galand, C., 'Perfect channel spliting by use of interpolation, decimation, tree decomposition techniques' *Intl. Conf. on Information Sciences/Systems,* Patras (August 1976).
28. Capellini, V., 'Digital filtering with sampled signal spectrum shift', *Proc. IEEE, 57,* 241–242 (1969).
29. Morris, L.R., 'Automatic generation of time efficient digital signal processing software', *IEEE Trans. Acoust., Speech, Signal Processing, ASSP–25,* 74–78 (1977).
30. McCabe, M.M., McCabe, A.P.H., Arambepola, B., Robinson, I.N., and Corry, A.G., 'New algorithms and architectures for VLSI' *GEC J. Sci. Technol., 48,* 68–75.

Professor Peter Burrough

Since 1976 Peter Burrough has been a faculty member of the Agricultural University, Wageningen. At the end of next month he leaves Wageningen to become Professor of Landscape Studies at the Department of Physical Geography, State University of Utrecht.

After obtaining his BSc, majoring in Chemistry, he was awarded the PhD from Oxford for his innovative work on statistical methods and computers in soil survey. This was followed by faculty work at Oxford, a tour with the Ministry of Overseas Development (spent mostly in Borneo), and Lecturer in Soil Science and Physical Geography at the University of New South Wales, Australia (1973–76) during which he was a member of the Australian Working Party on Soil Science, and guest lecturer at the Sydney University School of Tropical Medicine.

In recent years he has lectured at the International Institute for Land Reclamation and Irrigation, University of Novi Sad, Yugoslavia, and the Universities of Amsterdam and Utrecht, and has consulting experience in Norway, Britain, and Venezuela. He has pioneered computer graphics applications, including fractals, in soil science and physical geography, and the development of crop yield data bases.

He is Secretary of the International Society of Soil Science Working Group on Soil Information Systems, and a member of the Netherlands Steering Committee on Land Evaluation in the EEC.

Fakes, facsimiles and facts: fractal models of geophysical phenomena

Many geophysical phenomena (in which are included the form of coastlines, the topography of land, the distributions of sizes of islands and lakes, the discharges of rivers, the variation of climate or the variation of the properties of soil and rocks) appear at first sight to have no single characteristic length, area or time scale of variation. After analysing many data from maps, historical hydrological records and other sources, Mandelbrot[1,2] has concluded that the variation of many geophysical phenomena can be thought of as resulting from a particular stochastic, Brownian fractal process, characterized by a particular value of the fractal dimension D. These works present an overview of some 15 years of study. Mandelbrot has backed up his conclusions by using experimentally observed values of D as a controlling parameter in one-dimensional and two-dimensional simulations of landscapes that are not only remarkably striking, but which have a strong likeness to certain types of real landforms.[2] Although geographers and geoscientists have had ample time to examine and test Mandelbrot's ideas since they were first published in the late 1960s and 1970s, there seems to have been little work done either to confirm or to reject his claims. This paper presents a review of several recently published studies and attempts to examine just how generally applicable Mandelbrot's ideas are, where there are important deviations from the Brownian fractal models, and what the possible pitfalls and benefits are.

Basic properties of fractals

Mandelbrot[1] introduced the term 'fractal' specifically for temporal or spatial phenomena that are continuous but not differentiable, and that exhibit partial correlations over many scales. The term fractal, strictly defined, refers to a series in which the Hausdorff–Besicovitch dimension exceeds the topological dimension. A continuous series, such as a polynomial, is differentiable because it can be split up into an infinite number of straight lines. A non-differentiable series cannot be so resolved because every attempt to split it up into smaller parts results in the resolution of still more 'wiggliness'. For a linear fractal function, the Hausdorf-Besicovitch (or fractal) dimension *D* may vary between 1 (completely differentiable) and 2 (so irregular that it effectively takes up the whole of a two-dimensional space). For surfaces, the corresponding range of *D* lies between 2 (absolutely smooth) and 3 (infinitely crumpled). Table 1 presents some estimated *D* values for some geophysical data. In one topological dimension, landforms, groundwater levels and river flows generally have *D* values in the range 1.1–1.5 whereas soil and geological properties have *D* values in the range 1.5–2.0.

The second important characteristic of fractals is 'self-similarity'; when examined at increasingly higher resolution increasing amounts of detail are revealed that are scaled versions of the variations seen at lower resolution. For the stochastic fractals used in geophysical studies, the self-similarity refers to the statistical properties of the series and not to the exact shapes or forms. This idea implies that for landforms, for example, variations seen over a few metres should be statistically similar to those seen over hundreds or thousands of metres when transformed by a simple scaling parameter.

The concepts of self-similarity and characteristic scales in geographical phenomena have received a certain amount of attention in the literature. Scheidegger,[3] one of the few geographers to react to Mandelbrot's ideas, rejected the concept of self-similarity as absurd; Zonneveld[4] pointed out how when examining geomorphological features on aerial photographs

Table 1 *Estimated* D *values for some geophysical data**

Description	*Apparent* D	*Ref.*
Varve data	1.02 – 1.5	10
Tree ring indices	1.22 – 1.45	10
Annual precipitation	1.13 – 1.36	10
River flows	1.02 – 1.52	10
Thickness and lithology of Palaeozoic sediments	1.25 – 1.45	10
Iron ore in rocks	1.4 – 1.9	20
Groundwater levels	1.3 – 1.8	20
Landform	2.1 – 2.3	1
Soil properties	1.5 – 2.0	20,21
Landform	2.2 – 2.75	25

*A value between 1 and 2 represents an estimate from a transect or time series; a value between 2 and 3 represents an estimate from an areal survey.

their real scale could not be deduced from form alone. More recently, Church and Mark[5] have stressed that size and form are inextricably connected through the function and scale of geomorphological systems. Miller[6] discusses how the recent development and application of scaling methods to soil–water phenomena has allowed conceptual and analytical breakthroughs to be made in soil physics. Both Zonneveld's work and the work reviewed by Miller have proceeded independently of Mandelbrot.

Stochastic fractals—Brownian fractal function

With the possible exception of crystal structures, most geophysical phenomena are highly irregular, stochastic features that cannot be described by smooth polynomials or regular figures. Mandelbrot[1,2] and Mandelbrot and Wallis[7] have shown that a family of Brownian fractional noises (the fBm) can be used to model the stochastic properties of geophysical phenomena. Brownian linear motion (otherwise known as the Bachelier–Wiener–Levy process) is the simplest example of a random process having the properties of non-stationarity, a variation which is proportional to the length of the series, and statistical self-similarity. For a series of equally spaced spatial or time intervals, the sequence of increments of differences in the value of the property $Z(x)$, $Z(x+\varepsilon) - Z(x)$ is a normally-distributed Gaussian process with zero mean and variance σ^2. Brownian motion is said to be a statistically self-similar process in the sense that if the intervals are divided by a positive, but arbitrary ratio r and rescaled in the ratio r^H, where $H = 0.5$, the rescaled function has a probability distribution that is identical to that of the original function.

The semivariance (one-half the variance of increments) of the Brownian line function is

$$\gamma(h) = \tfrac{1}{2}E\left[Z(x+h) - Z(x)\right]^2 = h^{2H} \tag{1}$$

where $Z(x)$, $Z(x+h)$ are the values of the function at points x, $x+h$ separated by Euclidean distance h, and H is the scaling power parameter, here equal to 0.5. By allowing H to vary between 0 and 1, Mandelbrot showed that a whole family of statistically self-similar variations could be produced that cover the range of spatial intricacy from smooth curves to plane-filling noise. These series can be simulated by replacing the normally distributed Gaussian error term ε in the equation for Brownian motion

$$Z(x+\varepsilon) = Z(x) + \varepsilon \tag{2}$$

by a discrete fractional Gaussian noise (dfGn). These are stationary Gaussian random processes normalized to zero mean and unit variance defined as having the covariance

$$C(h.H) = \tfrac{1}{2}\left[|h+1|^{2H} - 2|h|^{2H} + |h-1|^{2H}\right] \tag{3}$$

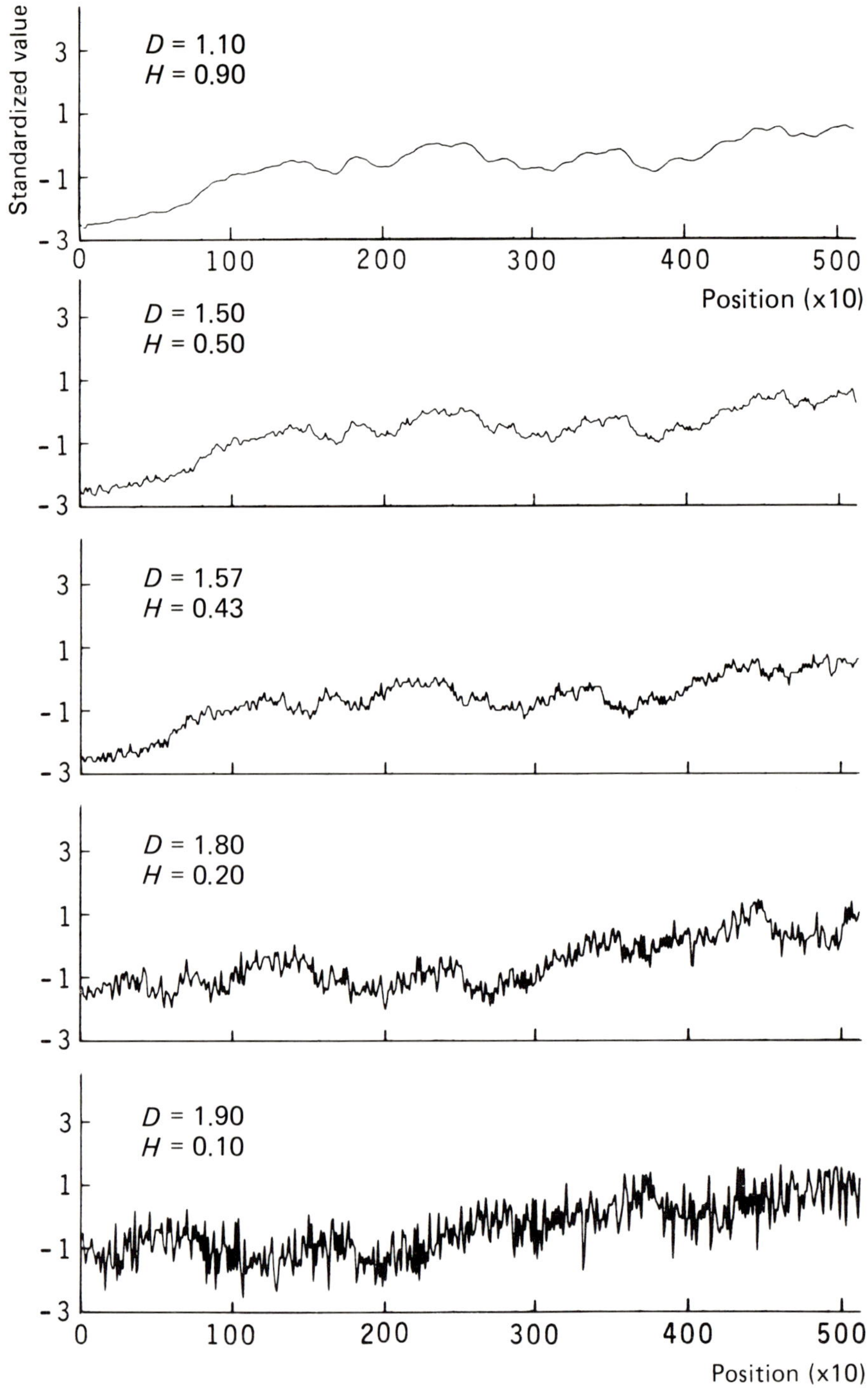

Figure 1 Variation of the form of the Brownian fractal function with D and H.

where h is the sampling interval, and $0<H<1$ is the scaling parameter as before. When $H>0.5$ the fractional noise has a positive correlation over the series, expressing the persistence of positive or negative values. When $H = 0.5$ or zero, the covariance is always zero and the fractional noise is identical to the Gaussian normal distribution of uncorrelated values. When $0<H<0.5$, the correlation between the 'past' and 'future' values of the noise is negative, indicating that positive and negative values tend to alternate.

When these noises are used to generate Brownian fractal lines, the positive or negative autocorrelation in the noise controls the form of the resulting functions. Large values of H (>0.5) result in smoothly varying functions in which long-term fluctuations are dominant; small values of H (<0.5) result in functions in which short-term variations are dominant (Figure 1). Mandelbrot[1,2] has described the two-dimensional analogues in great detail.

Estimating the fractal dimension

If an experimentally-observed series can be modelled successfully by a Brownian fractal function, then it should have statistical properties that are the same as those of the fBm. In particular, the D value should be consistent over a wide range of scales, and different methods of estimating D from the same data should yield comparable results. Table 2 lists six of the most important methods for estimating D that can be used for geophysical data, and which have been reported in the literature of fractal studies in geophysics and related fields.

Geophysical phenomena

The application of fractal analysis to hydrological time series using R/S analysis in the context of the Hurst phenomenon[8] has been thoroughly discussed in the literature[1,2,7,9–11] so I shall not cover the same ground here. It should be pointed out that there does not appear to be a universal acceptance that the Brownian fractal model is the only model that can be chosen to give an adequate explanation of the Hurst effect.[11,12] Doubts have also been cast on the quality of the data Hurst used and suggestions have been made that they were "distorted by long-term variability which is non-climatic in origin".[13] In particular, it has been suggested that the long-term variations could have been caused by simple changes in the level of means as a result of re-establishing or changing monitoring stations. Rather than reopen the discussion about how the Hurst effect should be interpreted, I prefer to consider some work done by various scientists in other areas of geophysical and geographical research to see how their results fit in with the Brownian fractal theory.

Rain and cloud areas

Lovejoy[14] used the area/perimeter relation (Table 2) to investigate the geometry of satellite- and radar-determined cloud and rain areas between 1 and 1.2×10^6 km^2. The data were well fit by the formula $A = P^{2/D}$, where $D = 1.35$, which suggested very strongly that no particular length scale was present. In climatological terms, this was a remarkable result. Extensions of the study down to 10m resolution, using measurements from aircraft, suggested that the relation held over an eight-power range of scale (10^2–10^8m). The implications of the study are that the study of turbulent

Table 2 *Summary of methods used to estimate* D *for geophysical data*

Name of method	*Relation*	*Estimate of* D
1. Length of a trail	$L(\lambda)=k\,\lambda^{1-D}$ λ is step size k is constant	Plot log L against log λ slope of line is $1 - D$
2. Area/perimeter relationship	$A = k\,P^{2/D}$ A is estimated area P is estimated perimeter k is a constant	Plot log A against log P slope is $2/D$
3. Variance of increments or semivariance (1-dimensional transect)	$2\,\gamma(h) = h^{(4-2D)}$ where h is sampling interval	Plot log $\gamma(h)$ against log h slope is $(4 - 2D)$
4. Power spectrum (1-dimensional transect)	$P(\omega) \approx \omega^{-(5-2D)}$ where $P(\omega)$ is the power and ω is frequency	Plot log $P(\omega)$ against logω slope is $-(5 - 2D)$
5. Rescaled range R/S analysis	The ratio of the rescaled range $R(t,s)$ for a series over time t, to the sequential variance $S^2(t,s)$ is proportional to the lag s.	Plot log R/S against log s Graph has a slope of $(2 - D)$
6. Korcak empirical relation for islands	$\mathrm{Nr}\,(A > a) = F'\,a^{-(D/2)}$ where $\mathrm{Nr}\,(A > a)$ is the number of islands above size a, F' is a positive constant.	Plot log $\mathrm{Nr}(A > a)$ against log a; slope is $-(D/2)$

Full definitions of these methods are given in refs 1,2.

isobars (which also have a theoretical D of ~1.35) should be able to extend our understanding of the distribution of clouds.

Impact craters on Mars

Woronow[15] was concerned with the problem of describing and comparing the shapes of impact craters of different sizes as seen on space-probe images of the planet Mars. Essentially, the problem was one of classification and description of regular features that can be much more easily related to Koch snowflakes than pits in a Brownian surface. Three types of impact crater had been recognized—pedestal, lobate and multilobate. The problem was to decide whether these were really different crater types, resulting from different processes, or whether they only appeared to be different as a result of the scaling and resolution limits of the imagery. The use of different imagery and resolution for measurement had caused apparent inconsistencies which impeded description, numerical taxonomy and modelling of these 'landforms'.

Using the relation

$$\frac{P^{1/D}}{A^{1/2} = K} \quad (4)$$

(a version of the area/perimeter method), where K is a constant, Woronow showed that three main populations of D/K relation occurred, and that each population corresponded with craters that could be termed pedestal, lobate and multilobate. He concluded that "multilobate ejecta deposits are not simply scaled-up versions of lobate deposits but that different physical processes or conditions control the ornamentation of their distal edges". In other words, Woronow used the estimates of D to suggest that the different forms resulted from different physical processes. By so doing, he has been one of the first authors to use the estimated D value as an analytical tool.

Shorelines, contour envelopes and lake outlines

Goodchild[16] studied the area/perimeter and length/scale relations of the coastline, the 250-foot contour and the lake outlines of Random Island, Newfoundland. This is a mountainous, glacially-affected island off the east coast of Canada. Data were taken from a 1:50,000 printed topographical map, digitized to an accuracy of 0.01 inch, which therefore gives a cutoff, or limit to the resolution of the study. All lakes of area greater than 0.01 square inches were included (153 lakes). Goodchild showed that the length/step size estimates resulted in relationships of the type $L(\lambda) = k\lambda^{1-D}$ for the shoreline and both contour sets, but not for the lakes which gave a sigmoid curve on the log/log pilot. In other words, the estimates of D for the lake outlines varied over the range of step sizes used. Grid-cell chain counts were also used to estimate *D*; the results are given in Table 3. Goodchild also estimated *D* from area/perimeter relationships and showed that for the contour envelopes and the lake outlines that these were inconsistent with the length/step size estimates. All estimates of *D* showed a consistent increase with altitude, however.

Goodchild also compared the form of the relation between rank and area for the contour envelopes and the lake outlines. Mandelbrot[1] has argued that the contour loops on a fractional Brownian surface should show strong-form Pareto or hyperbolic distributions (to which the Korcak empirical law is related), and has speculated that lake populations might also do so. The results for Random Island showed that the plots for log(rank) versus log(area) (log $Nr(A>a)$ versus log a) were *not* straight for contour loops nor for the lakes. The estimates of *D* varied over the part of the plot used to estimate them and they were not consistent with the estimates made from length/step size nor from area/perimeter methods. The distribution of lake areas fitted a log normal distribution better than the hyperbolic one.

Goodchild concluded that although various single aspects of the topography of Random Island exhibited the same type of behaviour as the fBm, the island as a whole differs in significant ways. Like Woronow, he pointed out that the increase in estimated *D* value with altitude was probably the result of different process. Coastal and fluvioglacial processes, which are dominant at lower altitudes tend to produce a smoothing in the relief compared with the frost-shattering and weathering processes operating at higher altitudes. Although there were many points where the topography of Random Island and an fBm model were similar, there were also so many points where the fBm model did not match that "no geographer would accept [the fBm model] as a new piece of the coast of Newfoundland".

Table 3 *Estimates of* D *for four line sets from Random Island*

Method used	*Shoreline*	*250 ft contour*	*500 ft contour*	*Lake outlines*
Length of trail (dividers)	1.11	1.19	1.31	1.3 – 1.8
Length of trail (cell counting)	1.13	1.33	1.54	1.61
Area/perimeter	1.14	1.17	1.19	1.30

Estimates are adapted from those given in ref. 16.

Variation of properties of soil and geology

The variation of the properties of the soil (e.g. clay or sand content, chemical composition, the depth and thickness of various layers, etc.) is very often much 'noisier' than the variation of landform or hydrological variables. There are many reported cases where the variation (usually estimated as variances within blocks of a given size, or as semivariances) appears to vary linearly on a log–log graph. Figure 2 presents data for some soil properties for an alluvial area in Arizona (taken from ref. 17). Smith[18] also presented many data that appeared to demonstrate a power law relationship between the variance of wheat yields and the size of plot. The distribution of gold ore in South African mines also shows a log–log linear relation.[19] All these cases would seem to be, *a priori*, situations in which the fBm is a useful model of how variations appear to have no discrete length scale. The large short-range variation of these properties (which may be partially the result of estimation errors) causes these properties to have large values of D, usually between 1.6 and 1.9 for linear transects.[20,21]

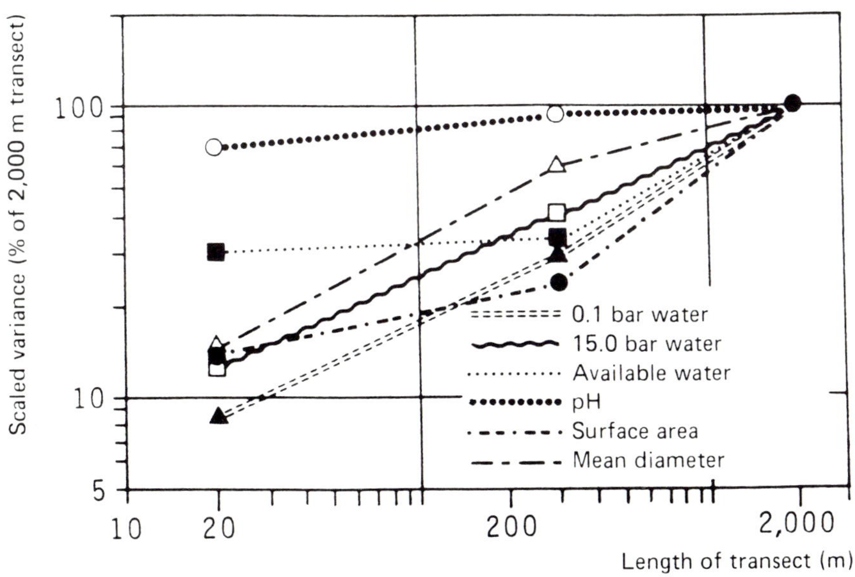

Figure 2 Variation of estimated variance with length of transect, using equal sample numbers but increasing sample spacing. Variance has been scaled as a percentage of the maximum observed over the whole sample area (data from ref. 16).

The power law relationships outlined above, are by no means the only kind of spatial variation observed for soil and geological variables, however; in fact, such log–log linearity seems to be the exception rather than the rule. Many experimental studies (see refs 19, 21–23) have found, that at least for defined areas of the Earth's surface (i.e. single sample areas) the semivariance does not increase without limit, but attains a more or less constant value at a certain range. This range is often, and with good reason, equated with the structure of the soil or geological phenomenon under study. Although Mandelbrot[2] has pointed out that "while our mind spontaneously analyses geomorphological configurations into superpositions of features having sharply distinct scales, these features need not be real", I would also like to add that they need not be wholly imaginary either!

Burrough[21,23,24] examined data from several soil transects to see if the fBm was a successful model. The data were taken from a 3,200-m long transect sampled every 10m in Oxfordshire (Sandford transect), from a 266-m long transect sampled every 1m across the experimental fields of the Proefstation voor Akkerbouw and Groenteteelt in de Volle grond, The Flevopolders, The Netherlands (PAGV), from a study of moisture supply in sand soils in the east of the Netherlands (Laren) and from a 1,060-m long transect sampled every 20m across the Maas terraces in Limburg, The Netherlands (Grubbenvorst). Figures 3 and 4 display a selection of the data and the semivariograms plotted on log–log axes. It can be seen that, particularly for the Sandford and Grubbenvorst data, there is a strong structure present that is related to the geological or geomorphological structure of the terrain. Later analysis[23] showed that the Laren data also had a clear structure that had been masked by the log–log plot shown in Figure 4: a linear plot of the data showed several clear knick-points in the semivariogram that could be related to a definite spatial structure. Figure 4 also shows that the estimates of D vary greatly with the part of the semivariogram chosen. Burrough[21,24] also examined the distribution of the increments along the series. For the fBm, these differences should by definition, be normally distributed Gaussian noise. Figure 5 shows that when a strong geological structure was present (the Sandford series for example), the distribution of the squared increments $[Z(x) - Z(x+h)]^2$ differed greatly from that of the fBm when the sampling interval was small enough to reveal the geological structure. When the sampling interval was too large to reveal the geological structure, the distribution of $[Z(x) - Z(x+h)]^2$ could not be distinguished from that of the fBm.

Burrough[23] also showed that the same data could be better fit by a set of nested, or superimposed variations, where each set of variation corresponded to a physical process operating at a discrete scale, and with a given weight. By using simulation models, he also showed that when short-range variations dominate (as they often do in soil) or when many independent sources of variation have been superimposed and no scale is dominant, then no appropriate model (fBm or nested, or other) can be unambiguously chosen.

A recent study of the fractal dimensions and characteristic scales of variation of landforms by Mark and Aronson[25] has also demonstrated effects similar to those observed by Burrough[21,23,24] for soil. Mark and Aronson estimated *D* for 17 digital elevation models for topographic quadrangles from the States of Pennsylvania and Oregon. They showed that in 15 of the 17 cases, the estimate of *D* was strongly dependent on scale. Moreover, the DEMs returned evidence for periodicity, with *D* values ranging from 2.2 to 2.75. Just as in Burrough's study, estimated *D* values increase with sample spacing.

Wind-blown sands

To my knowledge, no one has yet done a fractal study of dune morphology, although casual inspection of sand ripples and dunes on satellite images suggests a scaling process is at work.

In some situations, the wavelength of dune ripples is given by

$$\lambda = k/g\ (U + U_{crit})^2 \qquad (5)$$

where k is a constant, g is acceleration due to gravity, U_{crit} is the shear velocity and U is the wind velocity.[26] A visual examination of sand dunes on aircraft and satellite imagery and sand ripples on a beach suggests that the forms are just scaled versions of one another. However, wind-blown sands take up many other forms than simple ripples and dunes when the sedimentation processes are more complicated. For example, in Australia, there are many areas covered by parallel ribbon dunes many tens of kilometres long.

Historical records of climatic change

Climatologists frequently claim that patterns of climatic change are related to regular, deterministic cycles that can be linked to the motions of the Solar System.[27] A major problem seems to be that Fourier analyses of climatological records have resulted in estimates of cycles at all wavelengths from at least 2 to 200,000 years,[28] a situation that seems ripe for fractal analysis. Work on oxygen isotope ratios in ocean sediments reviewed by Covey[29] suggests that the periodicity of these ratios over a time span of 500,000 years can be largely attributed to 100,000-, 40,000- and 20,000-year cycles. These cycles have been linked to variations in the Earth's orbital parameters, the Milancovitch hypothesis. According to Covey "the most puzzling part of the problem . . . is why the 100,000-year cycle is stronger than the 40,000-year cycle" which in turn is stronger than

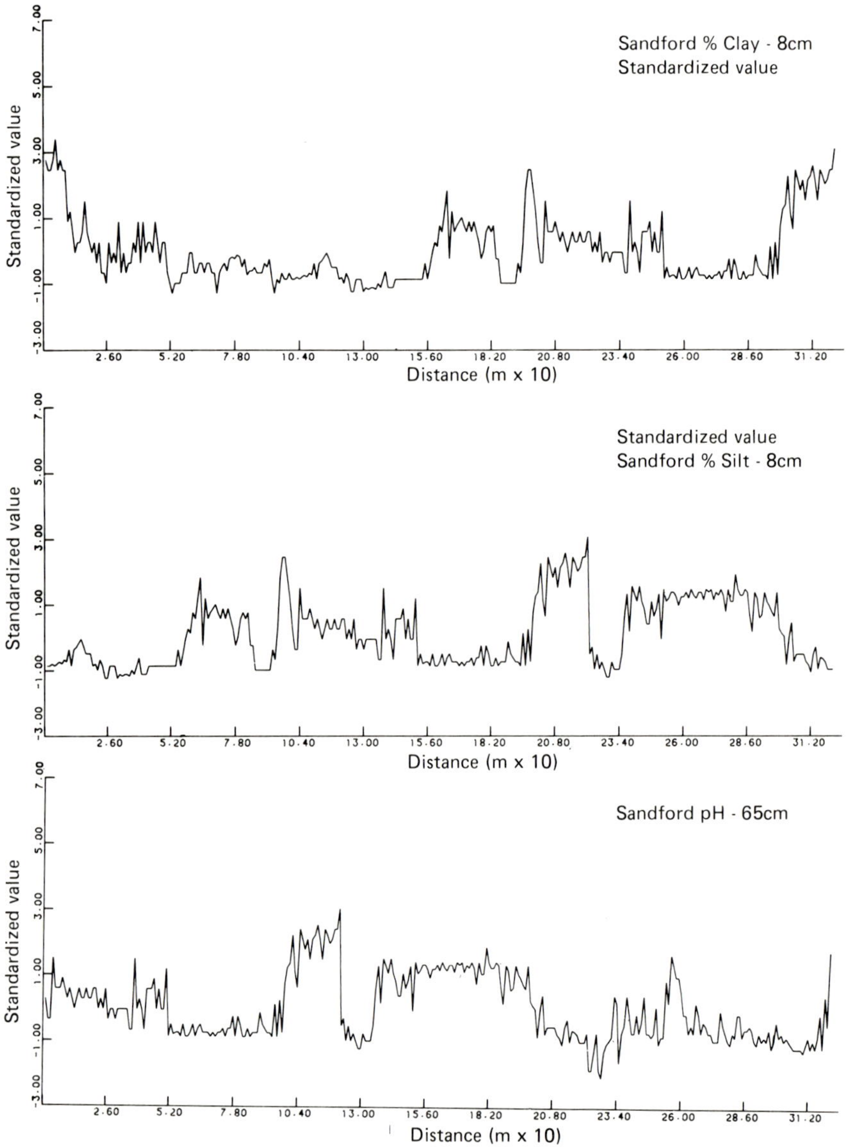

Figure 3 Plots of observed values of soil properties with distance. (All values have been scaled to zero mean and unit variance.)

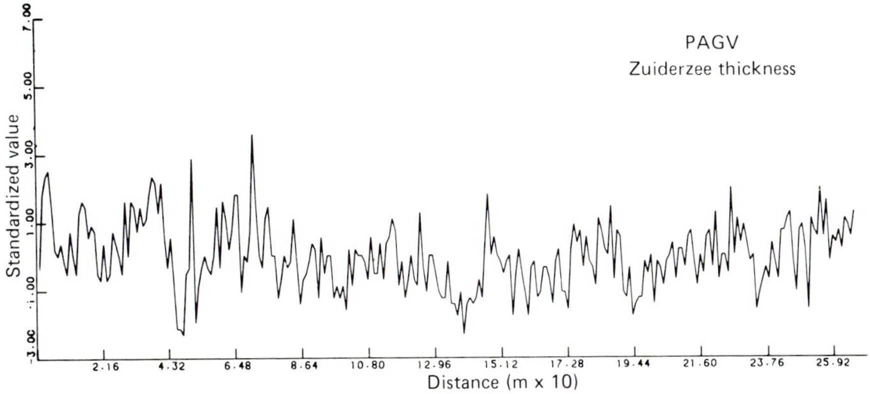
PAGV
Zuiderzee thickness
Standardized value
7.00
5.00
3.00
1.00
-1.00
-3.00
2.16
4.32
6.48
8.64
10.80
12.96
15.12
17.28
19.44
21.60
23.76
25.92
Distance (m x 10)

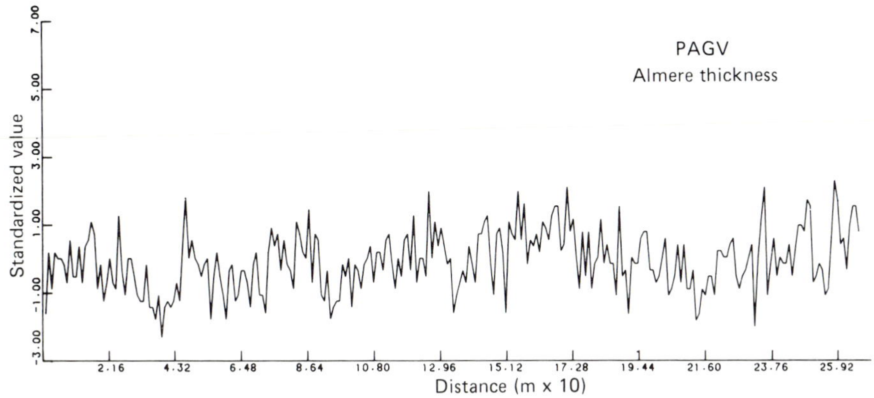
PAGV
Almere thickness
Standardized value
7.00
5.00
3.00
1.00
-1.00
-3.00
2.16
4.32
6.48
8.64
10.80
12.96
15.12
17.28
19.44
21.60
23.76
25.92
Distance (m x 10)

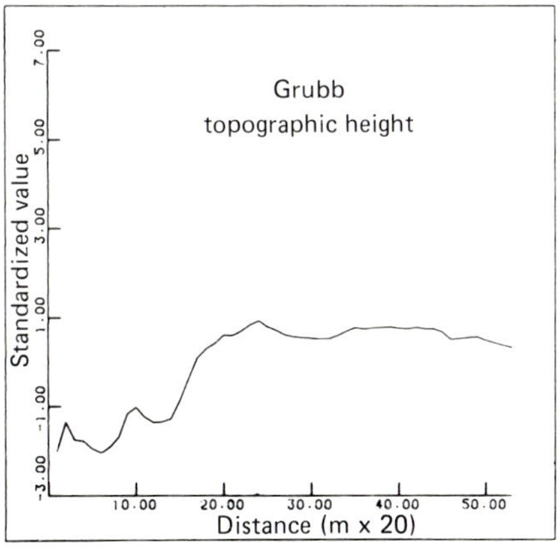
Grubb
topographic height
Standardized value
7.00
5.00
3.00
1.00
-1.00
-3.00
10.00
20.00
30.00
40.00
50.00
Distance (m x 20)

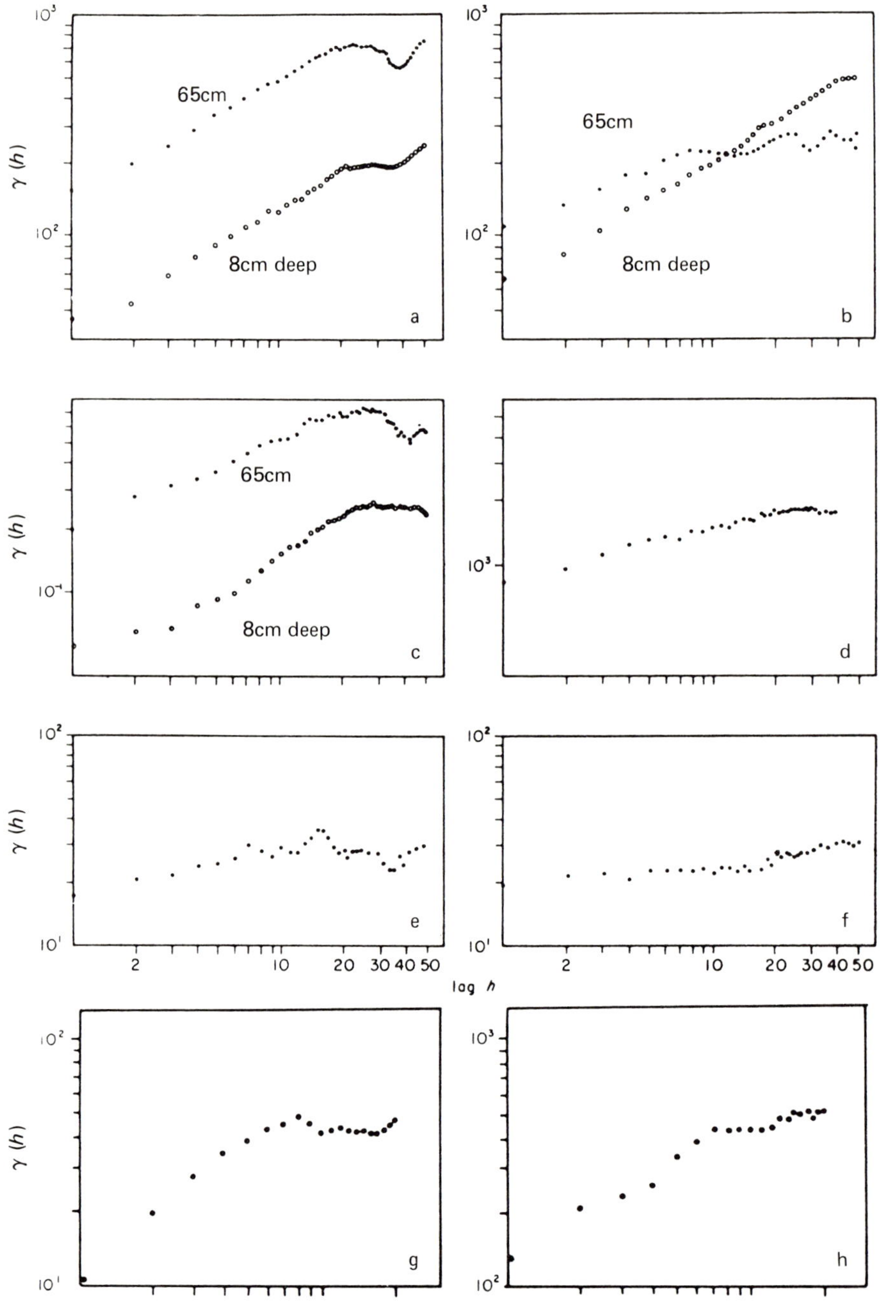
65cm
8cm deep
a
b
c
d
e
f
g
h
γ (h)
log h

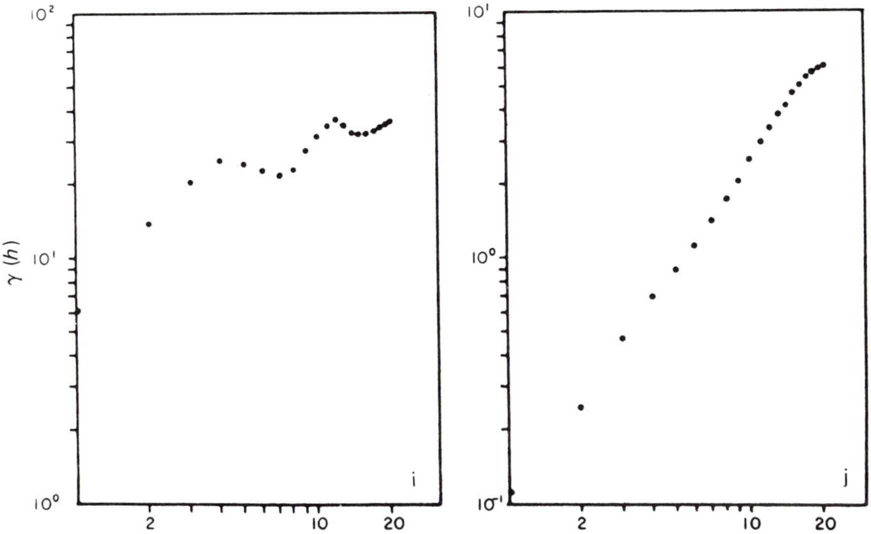

Figure 4 Log–log semivariograms of soil data computed to lag 50 (Sandford, PAGV), lag 40 (Laren) and lag 20 (Grubbenvorst). a, Sandford % clay; b, Sandford % silt; c, Sandford *p*H; d, Laren, estimated soil moisture capacity (mm); e, PAGV, thickness Zuiderzee sand (cm); f, PAGV, thickness Almere clay (cm); g, Grubbenvorst % clay (180–200cm) h, Grubbenvorst % clay (0–20cm); i, Grubbenvorst, thickness of the A1 horizon; j, Grubbenvorst topographic height (m).

"the 20,000-year cycle", a fact that is consistent with the power law behaviour of a fractal Brownian model.

More recent attempts to describe the vagaries of climate over periods of several hundred years combine cyclic effects with deterministic trends (the CO_2 concentration) and the supposed random outpourings of volcanic ash (see ref. 30). One aspect of climatic change that appears to be better fitted by the cyclic models than by a pure Brownian fractal is the often noted abrupt changes in mean temperature.

Discussion

The greatest problem of comparing the statistical properties of the real world with those of statistical models is that, whereas the models can be run hundreds of times with cheap, simulated data, real data are expensive to come by, and thus often limited in extent. Real data also include

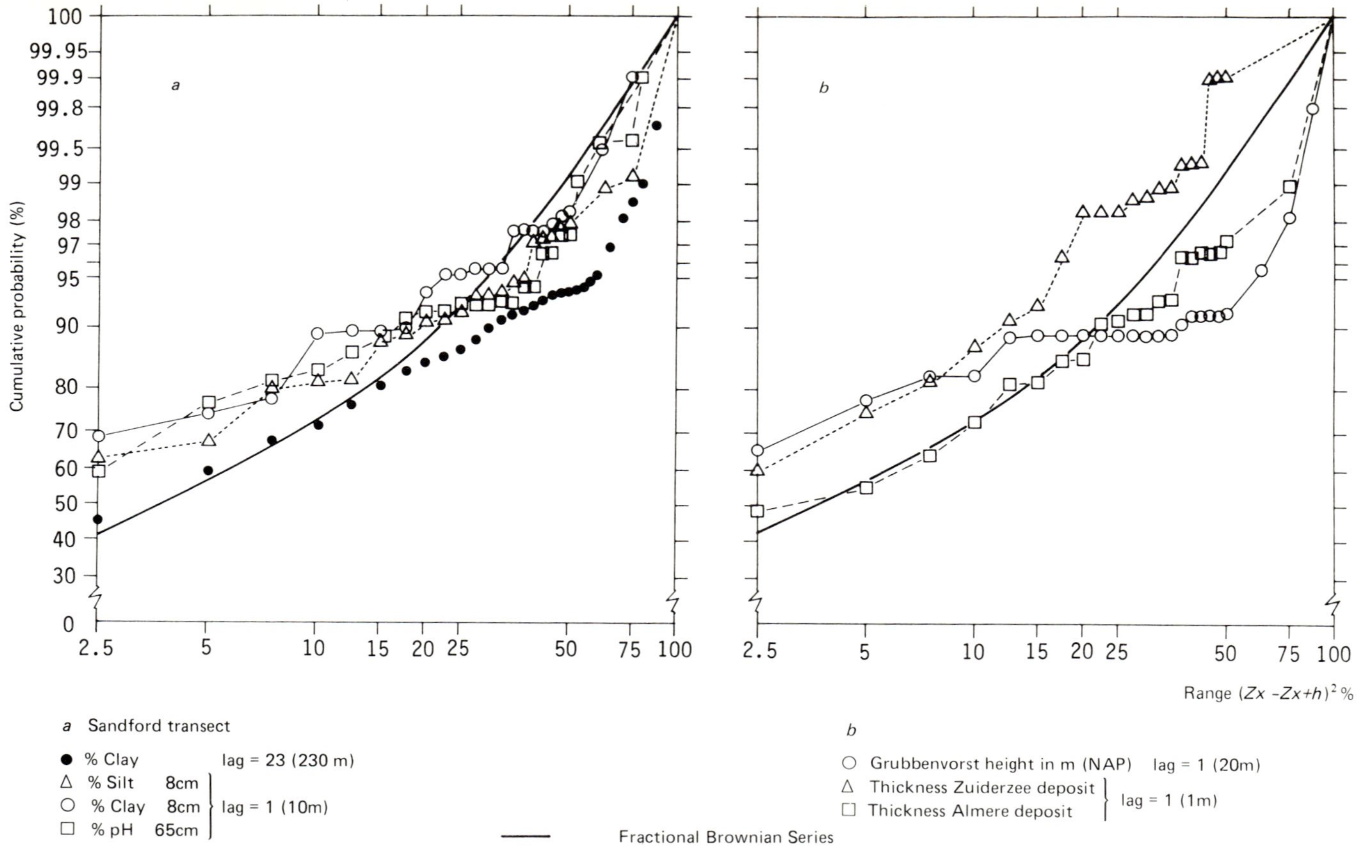

Figure 5 Plots of cumulative probability of the values of first increments squared for the soil data compared with fractional Brownian functions. Data with a definite spatial scale show strong deviations from the fractional Brownian model when the sample spacing is at an optimal level of resolution. When the sample spacing cannot resolve the structure the difference between a given experimental series and the fractional Brownian function is difficult to determine.

sampling variations, local anomalies, and measurement errors that can strongly affect our estimates and thus our interpretations and the choice of a "best-fit" model. With these limitations in mind, however, some generalizations can be tentatively made about the success of fractional Brownian models for geophysical phenomena. The fBm would appear to be an ideal model for phenomena that are largely the result of a single scalable geophysical process. For example, cloud cover appears to be controlled by a single turbulence process. Other examples are the scaling of soil–water phenomena and the form of sand dunes and ripples.

It seems also likely that the weathering of a homogenous mountain (one kind of geology) under a single climatic regime might lead to forms that could be described by the fBm and, if the weathered residue was bare of vegetation, self-similar forms may well be observed.

Most landscapes, and the sampled geophysical series obtained during investigation, are not the result of a single dominant process, but are the result of a complex interaction and superposition of processes. In any given local situation, however, a given process or material may dominate in a non-scalable way, and the resulting landform or geophysical data series may have a distinct and recognizable structure. For example, a geomorphologist usually has little difficulty in recognizing landforms produced as the result of tectonic uplift, be they in Borneo, New Zealand, the Alps or the Andes. Each single process may be reflected by its own fractional dimension, however, and just as in Woronow's martian study, perhaps we can use estimates of D to obtain greater insight through differences. In other situations, the combination of processes may have led to such complex interactions that no dominant 'structure' can be seen and no model can be unambiguously chosen. In still other situations, the observed variations may be the result of a combination of cyclic processes and random effects (for example, in historical climatology). The value of the fBm in geophysical studies does not come from pronouncements that this is the way the world should behave. Rather, it has considerable value as a null hypothesis, a point of reference against which the variations observed in particular sets of real world data can be compared.[15] The observation that D is constant over many scales may suggest that a single, scalable process is dominant; the observation that D changes with scale, or with altitude may indicate the dominance of different processes. It is disquieting that different methods of estimating D from the same data appear to give different results. Perhaps these differences in estimates of D are proportionately greater to differences between the real data and the fBm model, and may thus be a source of information about the scales over which dominant kinds of variation occur. If this can indeed be shown to be true, examination of the goodness of fit of a series to a Brownian fractal model could be a useful first step in exploratory data analysis (see ref. 31).

References

1. Mandelbrot, B.B., *Fractals, Form, Chance and Dimension*, Freeman, San Francisco (1977).
2. Mandelbrot, B.B., *The Fractal Geometry of Nature*, Freeman, San Francisco (1982).
3. Scheidegger, A.E., 'Stochastic models in hydrology', *Wat. Resour. Res., 6*, 750–755 (1970).
4. Zonneveld, J.I.S., 'Convergentie en luchtfoto-interpretatie. K.N.A.G., *Geograf. Tijdschr. VII* (1), 38–48 (1973).
5. Church, M., and Mark, D.M., 'On Size and Scale in geomorphology', *Prog. in Phys. Geogr., 4* 342–390.
6. Miller, E.E., 'Similitude and scaling of soil–water phenomena', in Hillel, D. (Editor), *Applications of Soil Physics*, Academic Press, New York (1980).
7. Mandelbrot, B.B., and Wallis, J.R., 'Computer experiments, with fractional Gaussian noises Parts I-III', *Wat. Resour. Res.*, 5, 228–267 (1969).
8. Hurst, H.E., 'Long term storage capacity of reservoirs', *Trans. Am. Soc. Civil Engng., 116*, 770–808 (1951).
9. Mandelbrot, B.B., and Van Ness, J.W., 'Fractional Brownian motions, fractional noises and applications', *J. Soc. Indust. Appl. Math., 10*, 422–437 (1968).
10. Mandelbrot, B.B., and Wallis, J.R., 'Some long-run properties of geophysical records', *Wat. Resour. Res., 5*, 321–340 (1969).
11. Klemes, V., 'The Hurst Phenomenon: a puzzle?', *Wat. Resour. Res., 10(4)*, 675–688 (1974).
12. Potter, K., 'Evidence for nonstationarity as a physical explanation of the Hurst phenomenon', *Wat. Resour. Res., 12*, 1047–1052 (1976).
13. Potter, K., 'Annual precipitation in the Northeast United States: Long memory, short memory or no memory?', *Wat. Resour. Res., 15*, 340–346 (1979).
14. Lovejoy, S., 'Area-perimeter relation for rain and cloud areas', *Science, 216*, 185–187 (1982).
15. Woronow, A., 'Morphometric consistency with the Hausdorff–Beiscovitch dimension', *Math. Geol., 13*, 201–216 (1981).
16. Goodchild, M.F., 'Randomness at Random Island: stochastic models of the physical landscape', *Paper presented to the A. Meet. Canadian Association of Geographers*, Corner Brook (August 1981).
17. Gajem, Y.M., Warrick, A.W., and Myers, D.E., 'Spatial dependence of a Typic Torrifluvent Soil', *Soil Sci. Soc. Am. J., 45*, 709–715 (1981).
18. Smith, H. Fairfield, 'An empirical law describing heterogeneity in the yield of agricultural crops', *J. agric. Sci., 28*, 1–23 (1938).
19. Journel, A.G., and Huijbregts, Ch.J., *Mining Geostatistics*, Academic Press, London (1978).
20. Burrough, P.A., 'Fractal dimensions of landscapes and other environmental data', *Nature, 294*, 240–242 (1981).
21. Burrough, P.A., 'Multiscale sources of spatial variation in soil. I. The application of fractal concepts to nested levels of soil variation', *J. Soil Sci., 34*, 577–597 (1983).

22. Burgess, T.M., and Webster, R., 'Optimal interpolation and isarithmic mapping of soil properties. I. The semivariogram and punctual kriging', *J. Soil Sci.*, *31*, 315–332 (1980).
23. Burrough, P.A., 'Multiscale sources of spatial variation in soil. II. A non-Brownian fractal model and its application in soil survey', *J. Soil Sci.*, *34*, 598–620 (1983).
24. Burrough, P.A., 'Problems of superimposed effects in the statistical study of the spatial variation of soil', *Agricult. Wat. Mangmt*, *6*/*2* 123–144 (1983).
25. Mark, D.M., and Aronson, P.B., 'Scale-dependent fractal dimensions of topographic surfaces: an empirical investigation with applications in Geomorphology and computer mapping', *J. Math. Geol.* (in press).
26. Allen, J.R., *Physical Processes of Sedimentation*, Allen and Unwin (1970).
27. Lamb, H.H., *Climate, History and the Modern World*, Methuen, London (1982).
28. Lamb, H.H., *Climate: Present, Past and Future—Climatic History and the Future*, Volume 2. Methuen, London (1977).
29. Covey, C., 'The Earth's orbit and the ice ages', *Scient. Am.*, *250* (2) 42–50 (1984).
30. Gilliland, R.L., 'Solar volcanic and CO2 forcing of recent climatic changes', *Climat. Ch.*, *4*, 111–131 (1982).
31. Tukey, J., *Exploratory Data Analysis*, Academic Press, New York (1977).

Professor Norman Dixon MBE

Norman Dixon is Professor of Psychology at University College London. After ten years with the Royal Engineers, during which he received the MBE for bomb disposal work in World War II, he returned to University to obtain a first in Psychology followed by a PhD, and a DSc from the University of London.

During his 28 years at University College London as Lecturer, Reader, and Professor he has researched extensively, particularly in the areas of subliminal perception, visually perceived movement, and military behaviour. He is the author of a number of books and papers on the theoretical aspects and practical applications of that work. Recognition of the special value of his researches includes the award of the Carpenter Medal for "experimental work of exceptional distinction", and Hon PhD from the University of Lund, Sweden.

Consciousness: the illusion of certainty

Since survival depends upon successful interactions with the environment, natural selection has favoured those animals best able to receive, store, and subsequently retrieve the most detailed representations of the external world.

But, since the evolution of a prodigious capacity for extracting information has resulted in the latter greatly exceeding the mechanical limits on what an animal can actually do at any one time, nature has also favoured those organisms with the greatest facility for acting selectively upon information received. Now, in man, the extraordinary and inexplicable product of this selective process is, of course, none other than his awareness of what has been selected. This 'having of conscious experience', a very late stage in our processing of information, confers three benefits. First, because of its very limited capacity, only very little—the highest-priority items of sensory inflow—are likely to be represented at any one time: a state of affairs that makes for speed of responding. Second, conscious experience constitutes a domain for emotions ('affects') and sensations; those feelings and signals that something is amiss and needs attention. Finally, the same mental apparatus serves as a sort of command centre for the planning and institution of adaptive behaviour.

All this is to the good, but unfortunately, one might almost say inevitably, a heavy price is paid for such a neat arrangement; namely an unwarranted and sometimes catastrophic confidence about the validity of what we consciously perceive. Indeed, it is just because this subjective picture of ourselves and our world is all we know—the apparently exclusive mediator between stimulus and response—that we jump to the sometimes fatal conclusion that the picture is necessarily correct.

Underlying such obvious manifestations of this 'illusion of certainty' as delusional thinking, non-lucid dreams (that is, when we do not know we are dreaming), and hallucinations that are taken for reality, are certain features of our psychobiology that, far from contributing to, actually threaten our survival. They include a capacity for stimulus-directed absent-minded behaviour, a lack of awareness of the actual precursors of conscious experience, a capacity for response to inputs of which the recipient remains unaware, and the functioning of emotionally-determined, involuntary defences against awareness of threatening stimuli. Much has been written about the experimental evidence for these phenomena, but one thing is certain—they necessitate a view of relationships between brain, mind, and behaviour very different from that traditionally held.

The apparent certainty of conscious experience could be mistaken in two ways, either through perceiving what is untrue, or through not perceiving what is true. Let us consider first the latter case.

Perception of truth

An inevitable consequence of the limitations that have been set upon our capacity for conscious experience is that the control of many habitual patterns of behaviour become relegated to brain processes which once set in motion, by some external stimulus, run their course without the need for conscious intervention. It is thanks to such sub-routines, which William James likened to "great flywheels", that consciousness can be left free to deal with other more pressing and interesting events. Most of the time this arrangement works tolerably well. But there are occasions when, through being preoccupied the conscious mind fails to note that the well-worn habit is for once quite inappropriate. The following example from the *Spectator* of 1711 (cited by Reason and Mycielska[1]) typifies such a happening:

> "My Friend Will Honeycomb is one of the Sort of Men who are very often absent in Conversation, and what the French call *à reveur* and *à distrait*. A little before our Club-time last Night, we were walking together in Somerset Garden, where Will had picked up a small Pebble of so odd a make, that he said he would present it to a Friend of his. After we had walked some time, I made a full stop with my Face towards West, which Will knowing to be my usual method of asking what/s a Clock, in an Afternoon, immediately pulled out his Watch, and told me we had seven Minutes good. We took a turn or two more, when, to my great Surprize, I saw him squirr (fling) away his Watch a considerable way into the Thames, and with a great Sedateness in his Looks put up the Pebble, he had before found, in his Fob. As I have naturally an Aversion to much Speaking, and do not love to be the Messenger of ill News, especially when it comes too late to be useful, I left him to be convinced of his Mistake in due time, and continued my Walk . . ."

From research[1] into the nature and causes of this phenomenon it seems that so-called absent minded slips are extremely common and may range in severity from such minor absurdities as saying "thank you" to a vending machine to such major lapses as that which caused the Moorgate tube disaster. From all the evidence, it seems fairly likely that in the latter instance the highly experienced driver of the ill-fated train was behaving in his approach to Moorgate *as if* he was on an earlier faster section of the line.

Unfortunately, such disastrous illusions of certainty are not unique to that particular catastrophe. Over the years a considerable number of mishaps on British railways, including the notorious Quintinshill accident of 1915 which, in the space of a few minutes destroyed five trains along with one entire battalion of soldiers, and the much more recent collision between three trains at Harrow and Wealdstone station, also with the loss of many lives, can be attributed to the misplaced reliance placed on conscious experience for a true and accurate picture of what is going on in the outside world.

Even more productive of such mental lapses is the extent to which the content of consciousness is determined by preconscious brain processes of which the recipient remains unaware. In the words of Popper[2] "all experience is interpreted by the nervous system a hundred-fold or a thousand-fold before it becomes conscious experience."

There are really three issues here of particular importance. Since the individual is unaware of his own brain processes he must also be ignorant of what these are doing to sensory inflow during that brief but measurable interval between the receipt of stimuli at peripheral receptors and the evoking of a conscious perceptual experience. Thus it is that we can be unwittingly deceived by a wide range of visual, haptic, and auditory illusions. Second, it is a feature of the relationship between conscious and preconscious processes that the individual cannot always know the extent to which perceptual experiences are drawing on memory rather than on current information from the external world.

As depicted in Figure 1 perceptual experience may be regarded as lying along a continuum from veridical and stimulus bound at one end to totally hallucinatory at the other.

Generally speaking as external stimuli get weaker and more ambiguous, and/or the individual becomes more sleep-deprived, emotional, or stressed, there is a tendency for perceptions to shift from left to right. The important point about this in the present context is not just the inability of

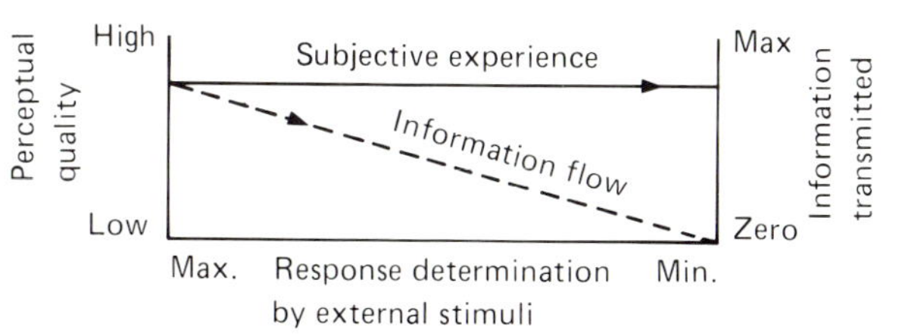

Figure 1 Diagrammatic representation of the relationship between information flow and subjective perceptual experience.

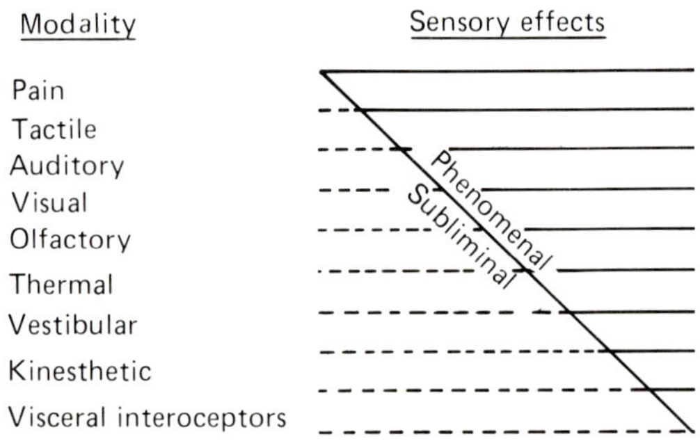

Figure 2 Hypothetical extents of subliminal and phenomenal effects for different sensory modalities.

knowing the extent to which perception is biased towards hallucination, but of always tending, when in doubt, to assume that what we see, hear or feel is a valid representation of the external world. Lucid dreams, for example (i.e. when we know we are dreaming), are extremely rare. Most dreams, however bizarre, are taken for reality—hence the relief (or disappointment) encountered on awakening!

Finally, as suggested in Figure 2, there may well be for all modalities a range of stimulus intensity between the physiological and awareness thresholds, within which stimulation may affect the individual without his knowledge. Not only is he unaware of being stimulated, but of the influence that this stimulation may be having upon him. Psychophysiological studies which have investigated this phenomenon support the view that rather than constituting a necessary stage in transmission of information through the brain, the processes responsible for conscious experience operate in parallel with those that are concerned with the transmission of information from input to output.

That there are two systems, as depicted in Figure 3, one for information transmission and the other for subjective mental life, each of which can enjoy a degree of autonomy and independence from the other, is evident from the fact that, on the one hand, we can have dreams and hallucinations in the absence of any sensory inflow and, on the other, be affected in a number of ways by stimuli which never reach awareness.

That the brain may respond to stimuli of which the conscious mind remains oblivious, has been most clearly demonstrated in experiments by Libet[3] and by Ádám.[4] In the Libet paradigm (see Figure 4) a stimulus to the hand was found to elicit the early components of the compound evoked

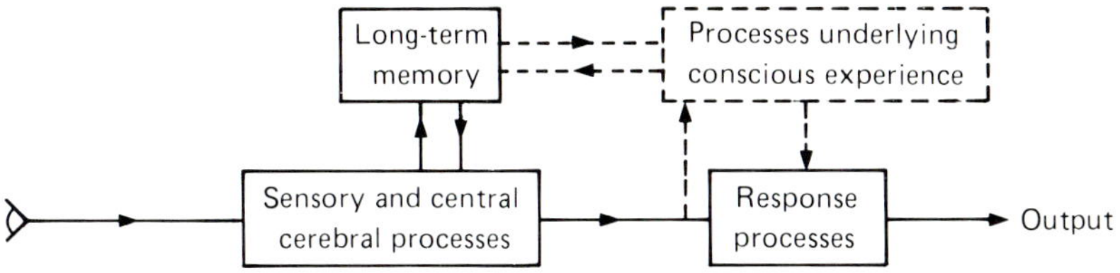

Figure 3 According to this diagram, incoming stimuli may initiate physiological effects leading to overt responses with or without any part of this traffic being represented in consciousness. Conversely, the content of consciousness may be drawn solely from memory and relatively, if not totally, be independent of sensory inflow (after ref.14).

potential (recorded, via burr holes in the skull, from electrodes on the somato-sensory cortex) even when its intensity was so low as to produce no awareness of its presence by the fully conscious subjects. With increasing intensity of the stimulus, reported awareness of the stimulus was accompanied by the appearance of later components of the C.E.P.S. Evidently the transformation of a purely physiological response, to one giving rise to a conscious experience occurred as a result of some major change in the brain's electrical activity brought about by an increase in the external stimulus intensity.

In the Ádám experiment (see Figure 5), cortical responses without any awareness on the subject's part were produced by the inflation of a balloon located in the duodenum. Whereas successive inflations of the balloon produced habituation (progressive cessation) of the evoked response, the inflation of a second balloon 15 cm from the first, resulted in immediate recovery of the previously extinguished response. In other words, unconscious brain processes can not only detect and respond to a subliminal stimulus but also differentiate between events at different locations.

A further contribution to the illusion that we are affected only by that of which we are conscious, comes from those defence mechanisms that serve to protect the conscious mind from unpleasant experiences. Two sorts of evidence are particularly pertinent to this contention—the symptoms of hysterical blindness and psychogenic deafness, whereby the patient for neurotic reasons simply fails to see or hear, and data from the numerous laboratory studies of perceptual defence. Typical of the latter was an early experiment by McGinnies[5] which showed that the conscious recognition of words or pictures depends upon their emotional significance. Investiga-

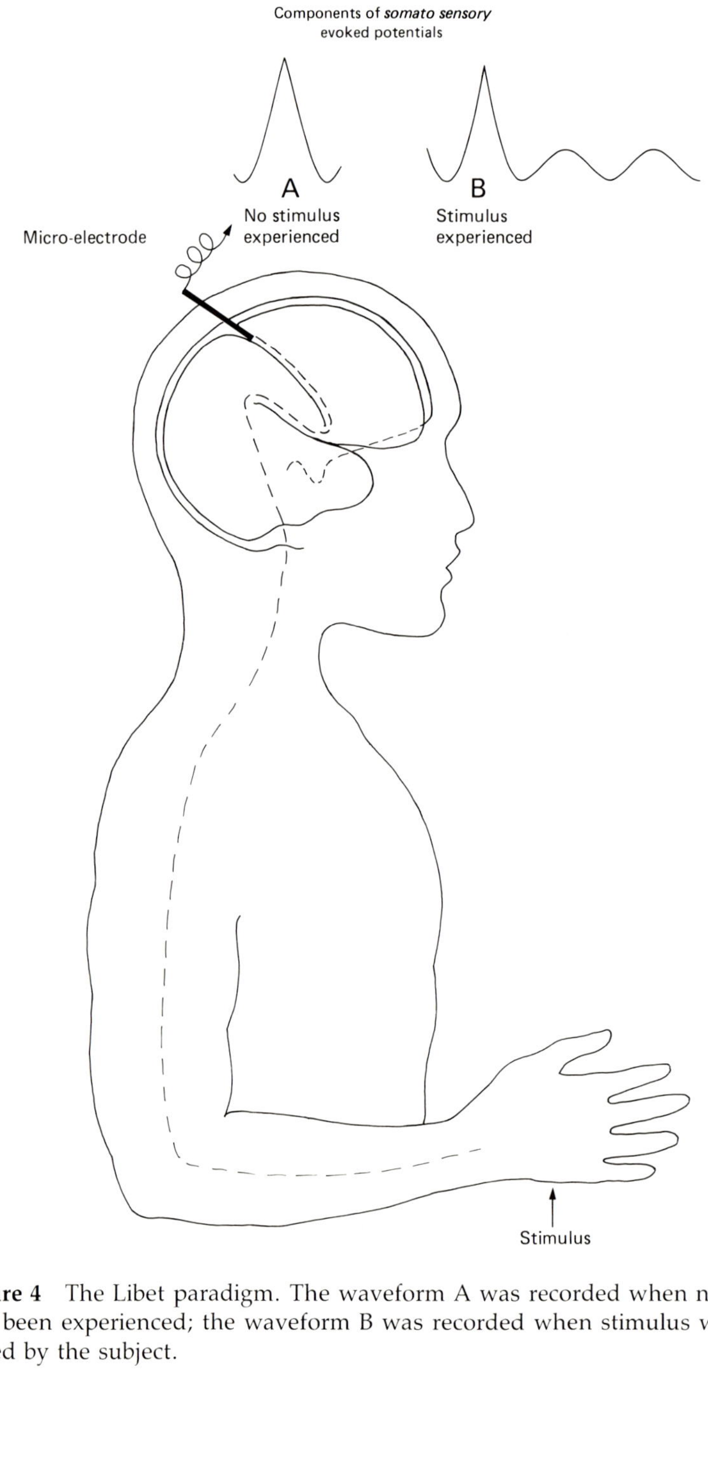

Figure 4 The Libet paradigm. The waveform A was recorded when no stimulus had been experienced; the waveform B was recorded when stimulus was experienced by the subject.

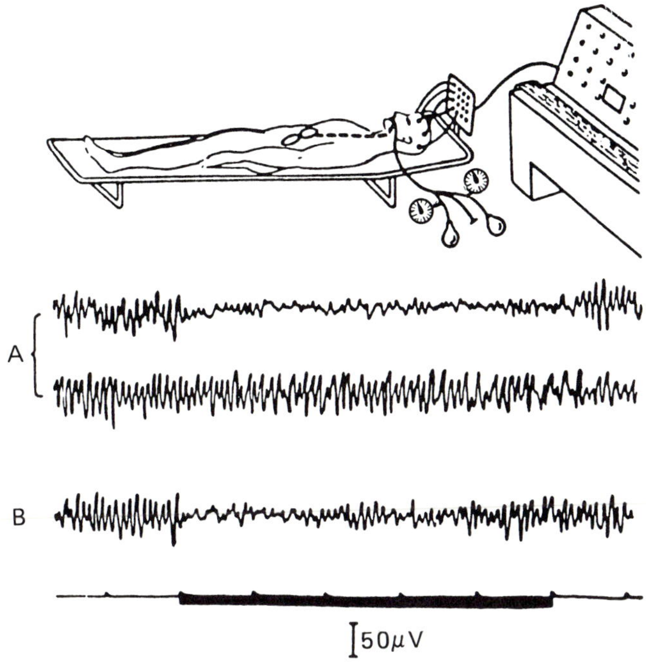

Figure 5 Duodenal discrimination tests in humans. Distension of balloon A caused desynchronization, which was soon habituated. Inflation of balloon B under such circumstances led to blocking of the alpha rhythm. Top: schematic representation of the experimental procedure. Bottom: time curve and duration of visceral stimulation, thick block.

tions of this phenomenon which involve either measuring changes in visual sensitivity of one eye when emotional words are presented below the conscious threshold to the other eye, or recording variations in the electrophysiological activity of the brain before awareness of threatening stimuli[6,7] suggest the following model of how the brain regulates sensory inflow as a function of the latter's emotional significance (see Figure 6).

According to this model, it is the preconscious cortical reaction to the meaning of sensory inflow which determines the contribution of the ascending reticular activating system upon which awareness depends. In other words, thanks to the fact that neural transmission is faster over the

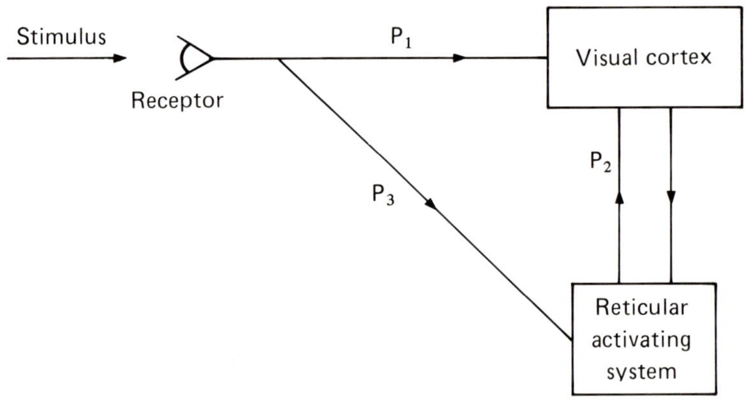

Figure 6 Possible physiological model for perceptual defence (after ref.7).

specific sensory pathway P_1 than over the non-specific collateral P_3 the brain can, without any reference to the conscious mind, determine what is and what is not suitable for conscious representation, or what is really affecting a person's behaviour, feelings, and bodily responses. Since we have a vested interest in the certainty of conscious experience, the whole notion of *unconscious* perception constitutes something of a threat to normal peace of mind. Consequently, the effects of unconscious perception have been rationalized (yet another defence which supports the illusion of certainty) as no more than watered down versions of normal conscious perception.

Let us consider a few of the many experiments which militate against this idea: First, there are studies by Walker[8,9] and Corteen *et al.*,[10,11] which support the apparent paradox that the brain may attach more importance to influences which are *not* consciously perceived.

In Walker's research, binocular rivalry is induced by presenting a red patch to one eye and a green one to the other. In this paradigm, subjects indicate, by pressing a key, whether they are seeing red or green. Once alternating rivalry has been achieved, the experimenter then adds a subliminal pattern stimulus to that colour patch which at that moment is directed to the 'blind' eye. Immediately that this happens, the subject reports alternation of conscious perception to the previously unseen colour.

There are two implications from this finding both of which make good biological sense. The first is that the brain is continuously monitoring sensory modalities for potentially important events and second that it is the hitherto unperceived of these which qualify for entry into awareness.

In an auditory counterpart of this experiment Corteen *et al.*'s subjects are asked to "shadow" (provide continuous report of) prose presented to one ear while individual words are presented at lower intensity to the other "unattended" ear. Consciously they hear the prose but not the unattended message stream. When threatening stimuli (i.e. words previously accompanied by an electric shock) are presented to the 'deaf' ear, they immediately manifest a galvanic skin response (recorded from the hand).

Since this result also held true for semantic associates to the shock-related words, we can draw the conclusion that without any interference of the conscious task (shadowing) the brain was able to analyse the meaning of unconsciously received stimuli and, on the grounds that galvanic skin responses are components of the emergency stress response, act appropriately in the interests of survival.

Particularly pertinent to the notion that unconscious perception is not a watered down version of normal perception, but qualitatively different is a recent study by Groeger.[12] People were asked to fill in the missing word in a sentence which was presented to one ear—"she looks __ in her new fur coat". For the completion they could choose between two words "smug" and "cosy". They performed this task under two conditions. In one case the word "snug" was presented to the other ear at a barely audible level, and in the other case, also on the unattended ear, well *below* the auditory awareness threshold. In the event it was found that the subjects chose "smug" as their completion in the first case and "cosy" in the second case. In other words, at the limits of *conscious* perception structure proved the deciding determinant but, for unconsciously received information, it was the latter's meaning that affected their choice of an appropriate completion.

One of the most convincing illustrations of how unconscious processes can mislead the conscious mind into a false sense of certainty comes from applications of a much used selection procedure, the Defence Mechanism Test.[13] In this test, the subject is asked to draw, and describe in words, the very briefly exposed picture of a young man. Whilst so doing he remains unaware of an "old ugly threatening male face" in the periphery of the test picture. As demonstrated in a 15-year validation study which followed up the careers of all those taking the test, this subliminal threat stimulus has a significantly disruptive effect on the drawings and descriptive behaviour of those who are subsequently found to be accident prone when learning to fly. Evidently, the unperceived threat activates defences against anxiety and, in so doing, interferes with the processing of information thus leading to errors.

Conclusion

So much for a very brief review of the extensive evidence for the fact that, far from providing faithful reflections of the external world, conscious experience gives, at best, a well edited outcome of complex interactions between many contributory influences, just one of which may emanate from somewhere outside our heads. One model which in encompassing these data gives a view of the relationship between brain, mind and behaviour, consistent with the present thesis, is shown in Figure 7 (see also ref.14).

When one takes into account the preconscious impact on sensory inflow of need states, emotional arousal, and the vast amount of information already stored in the brain's memory banks, and then acknowledges that

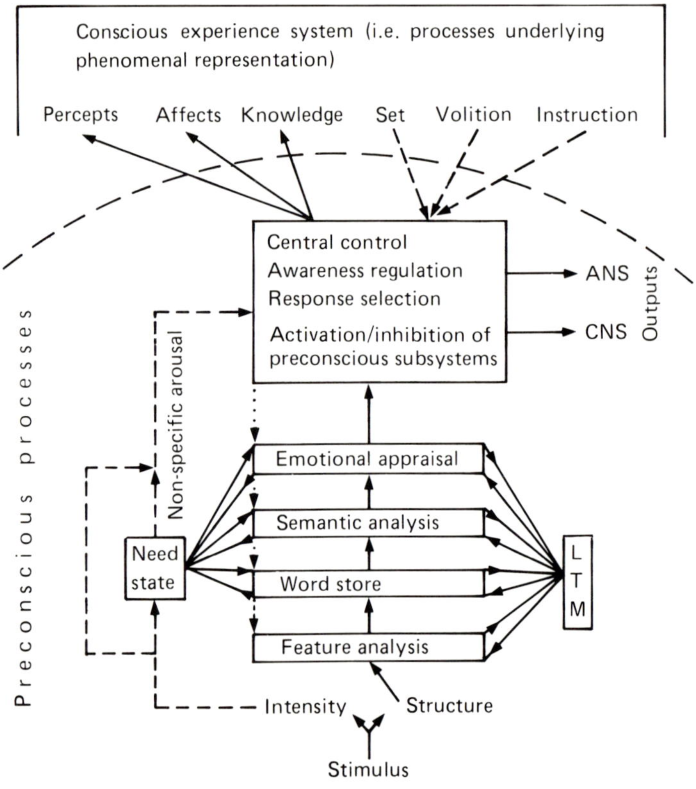

Figure 7 Relationships between conscious and preconscious processes implied by the data from studies of unconscious perception (after ref.14).

the limited capacity of consciousness may, without the owner's knowledge, be given over to feelings, imagery, and wish fulfilling fantasies which have nothing whatever to do with concurrent external reality, it should occasion no surprise that the apparent certainty of conscious experience is largely an illusion.

References

1. Reason, J., and Mycielska, K., *Absent-minded? The Psychology of Mental Lapses and Everyday Errors,* Prentice-Hall (Spectrum), Englewood Cliffs (1982).
2. Popper, K.R., and Eccles, J.C., *The Self and its Brain,* Springer, Berlin (1977).
3. Libet, B., 'Electrical stimulation of cortex in humans and conscious sensory aspects', in Iggo, E. (Editor) *Handbook of sensory Physiology,* Volume 2, pp.744–790. Springer, New York (1973).
4. Ádám, G., *Visceroception, awareness, and behaviour,* in Schwartz, G.E. and Shapiro, D. (Editors), *Consciousness and Self-regulation,* Volume 2, pp.199–213 Wiley, Chichester (1978).
5. McGinnies, E., 'Emotionality and perceptual defence', *Psychol., Rev., 56,* 244–251 (1949).
6. Dixon, N.F., 'Apparent changes in the visual threshold as a function of subliminal stimulation', *Q. J. Exp. Psychol., 10,* 211–219 (1958).
7. Dixon, N.F., and Lear, T.E., 'Electroencephalograph correlates of threshold regulation', *Nature, 198,* 870–872 (1963).
8. Walker, P., 'The subliminal perception of movement and the "suppression" in binocular rivalry', *Brit. J. Psychol., 66,* 347–356 (1975).
9. Walker, P., 'Binocular rivalry: central or peripheral selective processes?' *Psychol. Bull., 85,* 376–389 (1978).
10. Corteen, R.S., and Dunn, D., 'Shock associated words in a non-attended message: a test for momentary awareness', *J. Expl. Psychol., 102,* 1143–1144 (1989).
11. Corteen, R.S., and Wood, B., 'Autonomic responses to shock associated words', *J. Expl. Psychol., 94,* 308–313 (1972).
12. Groeger, J., 'Stimulus analysis: the function of level of level of reception', PhD thesis, Queen's University, Belfast (1984).
13. Kragh, U., 'Precognitive defensive organisation with threatening and non threatening peripheral stimuli', *Scand. J. Psychol., 3,* 65–68 (1962).
14. Dixon, N.F., *Preconscious Processing,* Wiley, Chicheter (1981).

Dr. Richard L. Garwin

Richard Garwin is an IBM Fellow at the Thomas J. Watson Research Center. He received a B.S. in Physics from Case Institute of Technology, Cleveland in 1947, a PhD in Physics from the University of Chicago in 1949, and after three years on the faculty of the University of Chicago joined the IBM Corporation in 1952. He is Adjunct Research Fellow in the Kennedy School of Government, Harvard University; Andrew W. White Professor-at-Large, Cornell University; and Adjunct Professor of Physics at Columbia University. He has been Director of the IBM Watson Laboratory, Director of Applied Research at the IBM Thomas J. Watson Research Center, and a member of the IBM Corporate Technical Committee.

He has made contributions to the design of nuclear weapons, in instruments and electronics for research in nuclear and low-temperature physics, in the establishment of the nonconservation of parity and the demonstration of some of its striking consequences, in communication systems, in the behaviour of solid helium, in the detection of gravitational radiation, and in military technology. He is author or co-author of some 200 publications including four books, and holder of 27 US patents. He has testified to many Congressional committees on matters involving national security, transportation, energy policy, and technology, and was a member of the President's Science Advisory Committee 1962–65 and 1969–72, and of the Defense Science Board 1966–69. Amongst many affiliations he is a Fellow of the American Physical Society, the American Academy of Arts and Sciences, and a member of the Council of the Institute of Strategic Studies (London).

Man, technology and uncertainty

This note provided a point of view on the topic of "Doctrines and Thinking on Nuclear War", as an introduction to discussion on this topic in the Working Group on the Strategic Arms Race at the Venice Pugwash meeting, August 1983. The evolution of technology, the growth in numbers of nuclear weapons, and the advent of new leaders on the political scene (as well as new people at these conferences) impels us once again to discuss these topics, no matter how distasteful and unsatisfactory it may be. My original subtitle, "Is There Wisdom Without Experience?" reflects the discrepancy between this field and those of experimental science, where fantastic theories can be put to the test and rejected (or accepted) according to experience. But none among us even hopes for a nuclear war in order to improve the quality of our theories in this field!

"Thinking About Nuclear War" is, in fact, too narrow a topic. We are concerned with nuclear *weapons*, and their acquisition as well as use. Nuclear war is an event of such magnitude that it cannot be considered in an existing environment—it changes the environment, and even the thinking in preparation for nuclear war has a major impact on the environment and on the likelihood of occurrence of that war.

So we might talk about *doctrine* for the use of nuclear weapons, *declaratory policy* for the use of nuclear weapons, and the *capability* to support this doctrine and policy. We must also talk about the nature of nuclear war. In a rational world of nations, doctrine and declaratory policy should at any time be consistent with capability, although one's aim should be that the three evolve together towards a more satisfactory and more secure posture. In a world of nation states, other nations are affected strongly by the nuclear weapon regime of a nation state, as is the entire international political and security environment.

My discussion owes much to my chapter in the book, *Nuclear Weapons and World Politics*.[1] I do not believe that we can or should attempt to eliminate nuclear weapons, so long as there is the possibility of war between nations. With the advance of technology over the past 40 years, and especially with the accumulation of stocks of fissile material from nuclear weapons programmes and from the civilian nuclear power programmes, even if all nuclear weapons were destroyed and their fissile material transferred to safeguarded stockpiles, it would be a matter of months, not years, before a nation locked in conventional combat would obtain nuclear weapons crude by today's standards, but very much better than those used in 1945. Thus the question is to find a regime to provide international security in the presence of nuclear weapons, and preferably one in which the existence of nuclear weapons provides greater security than would be the case in a world in which nuclear weapons could not exist.

History

Serious effort on the development of nuclear weapons began in Great Britain and the US early in World War II, culminating in the use of two 20-kiloton fission bombs on Hiroshima and Nagasaki in August 1945. The work was motivated initially by fear that Nazi Germany would develop such a weapon, but in view of the tens of millions of people already killed in World War II the bomb would surely have been used in 1945 against Germany as soon as it was available. Perhaps the opportunity for surrender would have preceded use, but there was hardly the opportunity to develop doctrine or policy.

After 1945, the US had a monopoly on nuclear weapons until 1949, and effectively for some time thereafter. Nuclear weapons were produced at a modest pace, but delivery by aircraft was not guaranteed of success. A monopoly on nuclear weapons (with some means of delivery) provides a nation not only *deterrence* of hostile acts by other nations, but also *compellence*—the ability to force another nation to accede to demands. The US did not take advantage of this power, but it was latent and unacceptable to other nations, especially the Soviet Union. Logically, the unacceptability of sole possession of nuclear weapons could have been countered by supranational possession, but international cooperation in the nuclear age was not yet developed to that state, and the opportunity was lost.

Nuclear monopoly allowed a strategy of 'massive retaliation', by which conventional attack of the other side against the nuclear power or its allies could be countered by devastating response against the homeland of the attacker. Even the first two nuclear weapons used in war had killed almost 100,000 persons each, so that relatively few fission bombs would suffice to render any potential gain from conventional aggression minimal by comparison, and so to deter conventional attack.

As fission bombs became more numerous, and as thermonuclear weapons of multi-megaton yield were established as feasible by the 'Mike' explosion of November 1952, one could conceive of defeating conventional armies in the field. Nuclear weapons were also deployed as warheads of otherwise commonplace antiaircraft missile systems, and became available as nuclear-armed torpedoes, land mines, and the like. The potential effectiveness of such weapons against traditional operation of conventional armies forced a rethinking of military tactics, so as to reduce the vulnerability of military forces to these nuclear weapons. Even though the armies thus adapted were considerably less vulnerable to nuclear weapons, the additional constraint made them less capable in their conventional role. Thus the threat of nuclear attack against conventional armies diminished the effectiveness of the other side. So long as one had a monopoly on nuclear weapons (in this case, tactical nuclear weapons), the mere possession of nuclear weapons improved one's (relative) conventional capability. However, it was the threat of use of the nuclear weapons which conveyed these benefits. With the potential tactical use of nuclear weapons by the other side, and with the advent of tactical nuclear weapons delivered by short-range rockets, one's own forces lost conventional effectiveness because of their need to adapt to the nuclear threat.

In fact, in the period of nuclear monopoly, the threat of use of nuclear weapons against the society would have served just as well to deter conventional attack. The capability to use nuclear weapons against military forces in the field was sought and achieved at great expense, in large part because the use of weapons against civilians is forbidden by the rules of war and is especially repugnant to professional military officers. The experience of World War II, however, shows that in large-scale conflict one side or the other is likely to ignore these restrictions on the use of effective weapons, under which circumstances any codification of the rules of war appears to allow retaliation in kind.

It was evident from the first, however, that the enormous effectiveness of nuclear weapons against human beings in cities limited the degree to which a nation armed with nuclear weapons could use those nuclear weapons in combat against the military forces of the other side. The other side could, if the stakes were high enough, use or threaten to use a nuclear weapon against a city, thereby destroying 0.5–1 million people with a single megaton weapon. Individuals on the battlefield are far more resistant to nuclear weapons than are houses and normal buildings (and the people within), but a more important factor is population density over the 100 square kilometers or so kill-area of a megaton burst. Thus began a race between strategic defense and the evolution of the strategic offensive forces, in order to prevent the emergence of an invulnerable homeland which would permit once again the use of raw force on the battlefield—this time including nuclear weapons.

Of course, a primary target of one's own military force (including nuclear weapons) is the effective military force of the other side—especially strategic nuclear weapons. Over the years, these strategic offensive forces have evolved so that they are not readily destroyed and so that they can penetrate existing defenses on the other side. They have evolved differently on the two sides—the US having most of its warheads on strategic submarines and an increasing number as cruise missiles on bombers, while the Soviet Union has concentrated 75% of its warheads on Inter-Continental Ballistic Missiles (ICBM). These disparities reflect differences in the geographical situation of the two nations and in the state of their technology.

Recent influences

Four problems continue to afflict 'deterrence by threat of retaliation':

—the moral quandary of holding innocent people hostage for the decisions of their leaders,
—the concern that if war does start accidentally or through misjudgment, one does not preserve one's own country by actually destroying the other's,
—that if war does start, one may have vastly more nuclear weapons and delivery systems than are required to destroy the other side, and would like to use some of them to reduce the damage to one's own country,
—that the threat of total destruction may not be deemed credible for "just a little aggression", or even for an attack on one's ally.

These various problems have been addressed in policies known as 'limited response', 'flexible response', 'extended deterrence', and the like. In fact, official statements indicate that the US does not explicitly target weapons on civilians, but on military forces and military-related industry. The fact that people are located close to such industry, however, means that there is very little difference in outcome between this policy and one that targets civilians specifically.

The Pastoral Letter of the American Catholic bishops,[2] "The Challenge of Peace: God's Promise and Our Response", wrestles with the apparent effectiveness of deterrence (of nuclear attack) by threat of retaliation against the society of the attacker, which would clearly result in the slaughter of mostly innocent people. The Bishops are led to a "strictly conditioned moral acceptance of nuclear deterrence", but assert that "'sufficiency' to deter is an adequate strategy; the quest for superiority must be rejected."

Another recent influence is the modernization and expansion of the Soviet intermediate-range nuclear force threat to Europe, with the deployment of the SS-20 mobile, MIRVed missile, and the planned response by NATO of deploying 464 ground-launched cruise missiles and 108 Pershing IIs beginning the end of 1983, absent an agreement which would persuade NATO that this deployment was unnecessary.

A third major influence is President Reagan's announcement 23 March, 1983 (and the clarification by Secretary of Defense Weinberger 27 March) of the adoption by the United States of a new goal of absolute defense against nuclear weapons, ". . . rendering these nuclear weapons impotent and obsolete." The President has direced a comprehensive "effort to define a long-term research and development programme to begin to achieve our ultimate goal of eliminating the threat posed by strategic nuclear missiles. This could pave the way for arms control measures to eliminate the weapons themselves." He notes, "And as we proceed we must remain constant in preserving the nuclear deterrent and in maintaining a solid capability for flexible response."

A fourth major influence is the report of the Presidential Commission on strategic forces (May 1983, headed by LGen Brent Scowcroft) which made the very sensible observations that vulnerability of the land-based ICBM component alone of the strategic forces was unimportant, in the context of overall force survivability; that the US would be more secure by developing a small, single-warhead ICBM as the future direction of the ICBM force; that small submarines should ultimately be available to augment or replace the large submarines in the SLBM force; and that the currency of arms control should be numbers of *warheads* rather than launchers, in order to control the threat directly. Contrary to its explicit rationale, however, the Scowcroft Commission unanimously recommended the deployment of 100 MX missiles in Minuteman silos, claiming that the small ICBM (Midgetman) would be available only in the 1990s. This observer believes that the Midgetman could be available by 1989 if it were not artificially delayed, as was the Trident-I (C-4) missile.

Politics of the MX

> "I can call spirits from the vasty deep." boasted Glendower in Shakespeare's Henry IV, Part I. ". . . so can any man, but will they come when you do call for them?" retorted Hotspur.

Just so—doctrine must make some contact with capability, and proclaimed goals with feasibility. Furthermore, although uncertainty, deception, and even pretence have their place in human affairs and international relations (including in deterring attack), self-delusion can hardly be beneficial.

Unfortunately, even technical alternatives towards the same strategic goals are often not considered with an open mind; how much more difficult it is to compare alternative strategies and doctrines! This seems the nature of human beings—exhibited in US and USSR alike; it is one of the personal dificiencies that good government solves and bad government exaggerates. The MX deployment proposed by the Scowcroft Commission May 1983, is a case in point. The members of the Commission quite frankly acknowledge that the MX in Minuteman silos will be just as vulnerable as Minuteman, and a more attractive target, but they maintain (as does President Reagan) that the deployment of MX is essential in order to advance the cause of arms control. Individual members of the Commission are more explicit in their testimony and statements and say that it is the threat of deploying far more than 100 MX (and so getting an apparent first-strike capability against Soviet ICBM silos) which will increase Soviet motivation to *reduce* numbers of large land-based missiles. Whatever the validity of the argument, the MX development and deployment comes at great cost—at least $15 billion, and the same silo-killing capability can be obtained sooner by improvement of accuracy of the existing Minuteman missiles at a cost substantially less than $6 billion. A greater (and earlier) first-strike capability against silos can be obtained by fitting Trident-I missiles with guidance systems which make use of Navstar satellite navigation signals, thereby providing almost 5,000 warheads of shorter flight time than ICBMs with a silo-killing capability.

But it is not at all clear that the USSR can be induced by threats to incur vast costs of uncertain benefit. Of course, it was unnecessary for purposes of deterrence for the USSR to build such large numbers of accurate MIRVs, and whatever US analysts think of the validity of a model of the arms race in which the Soviet Union responds to US initiatives, the Administration maintains quite frankly that it is building weapons and modernizing its forces in response to Soviet weapons.

Strategic defense

So the MX is unnecessary and excessively costly whatever US demands for a strategic posture! How about the desire for perfect defenses against nuclear weapons, presumably to be teamed with extensive civil defense, while still maintaining an offensive strategic capability?

The feasibility of a defence is conditional upon the requirements. If one side with 3,000 warheads sets itself the demand to destroy at least 1,400 of 1,500 hardened silos, it is quite easy to frustrate such an attack, since with reasonable missile reliabilities the defense need destroy or defend successfully against no more than one re-entry vehicle at each silo. However, if the purpose of the defense is to *ensure* that fewer than 20 re-entry vehicles of an initial force of 10,000 strike their city targets, the defense has a vastly more complicated task, even if it is not attacked, confused, or bypassed.

If there is any basis for President Reagan's initiative of 23 March 1983, it must lie in a ballistic missile defense (BMD) capable of destroying ICBMs and SLBMs during the boost phase, which for ICBMs is a long distance inside the USSR and far from bases of BMD. Furthermore, the boost phase for the MX lasts only 155 s., and even for beam weapons whose effects are carried with the speed of light, the time for assessment and command is critically short. But there are fundamental problems with such systems, independent of the weapon technology.

It seems to me that there is essentially no chance of preserving US society against a large nuclear strike by the USSR, or vice versa, no matter how hard we work on defenses which have been discussed thus far either openly or in government circles. Nor do such (imperfect) defenses significantly add to deterrence, already accomplished if one has the capability, the doctrine, and the will to use large numbers of nuclear weapons in retaliation for attack. At a level of a few nuclear weapons exchanged between the superpowers, defenses might be effective even in preventing nuclear detonations on cities, but 100 nuclear weapons could certainly be made available to destroy a city, making a grisly point without disarming the attacker.

Between the superpowers, imperfect defenses will simply lead to more rapid escalation if nuclear war begins. However, massive defenses *might* have some effectiveness against the strategic nuclear forces of a lesser power, especially if that power has no effective warning system to launch its vulnerable weapons before they are destroyed in their bases, so that only a fraction of the force need be handled by the defenses.

However, in addition to weighing the potential benefits of defenses if war comes, or of assessing the reduction in likelihood of war, one must consider the disadvantages—that destruction if war comes might be *increased* because of augmentation of the strategic offensive force, because of such augmentation and greater application to counter-value targets, together with attacks on the defense which totally defeat it; or because defenses on both sides make one side (perhaps rashly) confident that damage will be reduced, but worried that the other side will be able to improve its defense in a while so that it will be effective.

If one side believes large-scale nuclear war to be truly inevitable, then it is only the time when war begins which must be fixed, and that side is likely to fix it at a time of its own least disadvantage. It is unconditionally suicidal to persuade one's opponent that nuclear war is inevitable.

But if nuclear war comes, must we not have planned for the actual use of nuclear weapons—else they will be totally unusable and not serve to deter that war? Yes, but there are limits to the effort and the detail of that planning, just as there are limits to the numbers of nuclear weapons which need be available for retaliation. Indeed, those upon whom the society depends to launch those nuclear weapons must think about how they will do it, must find targets, must practice procedures; but the quality of deterrence does not depend in detail in the targeting of such vast numbers of nuclear weapons, at least judging by the values held by any human being I know, whether citizen or leader of the US, the USSR, or any other technologically-advanced nation.

Conclusion

Political and military leaders of the US and the USSR tell us of the expected unparalleled devastation of large-scale nuclear war. General David C. Jones, Chairman of the Joint Chiefs of Staff of the United States (retired 1982) expresses grave doubts that any nuclear war could be limited or protracted. In the US, at least, expansion of strategic nuclear weapons is supported by arguments that such expansion improves the 'quality of deterrence'—that is, that they affect the decisions on the other side. Would an American President be more deterred from initiating all-out nuclear war against the USSR by the thought that accurate Soviet MIRVs could destroy inaccurate US ICBMs (as was the argument in the late 1970s) or rather by the thought that inaccurate Soviet warheads would land on some US cities, and on *all* US cities if there were a response?

Would a Soviet leader be more 'deterred' from nuclear war by the thought that accurate US ICBMs could destroy half of the empty Soviet ICBM silos and thus prevent their being reloaded with spare missiles, or would he be just as deterred by inaccurate US ICBM warheads causing sufficient fallout at those silos to prevent reload? Or by the certainty that nuclear weapons will destroy Soviet society if Soviet weapons are used to destroy another's?

My view is that the US and the USSR would both be considerably more secure if they renounced the intention of destroying the opponent's strategic offensive force and began to improve the survivability of their own strategic offensive force by arms control agreements and by unilateral action, such as a move to smaller submarines, to somewhat fewer warheads in Midgetman silos, and in understanding the 'prisoner's dilemma' in which each side finds it in its short term interest not to cooperate or to reduce the threat to the other side, thereby insuring an increasing long-term peril to itself.

References

1. Gompert, D.C., Mandelbaum, M., Garwin, R.L., and Barton, J.H., *Nuclear Weapons and World Politics—Alternatives for the Future*, pp. 81–147, McGraw-Hill, New York (1977).
2. *The Challenge of Peace: God's Promise and Our Response*, United States Catholic Conference (1983).

Professor Geoffrey Eglinton

Geoff Eglinton is Professor of Organic Chemistry at Bristol University. He gained his PhD in Chemistry at Manchester in 1951, and afterwards did post-doctoral research at Ohio State University with Professor M. S. Newman and at Liverpool University with Professor A. Robertson. After a Sabbatical year at the University of California, Berkeley, he became Lecturer in Organic Chemistry at Glasgow University in 1954. His DSc was gained at Manchester in 1966. He went to Bristol University in 1967 as Senior Lecturer in Organic Geochemistry and Head of the Organic Geochemistry Unit, and in 1973 was appointed Professor.

His research interests span the study of the occurrence and distribution of organic compounds in rocks and fossils; environmental organic chemistry; and microanalytical techniques. The overall theme is the study of carbon chemistry from environmental, geological, and planetary aspects.

He is the author of well over two hundred scientific publications, a member of numerous national and international organisations, and he participates in scientific programmes on BBC television.

The alchemy of life: probing the molecular record

What record has life on this planet left molecularly—that is, as molecular or chemical fossils? There is abundant evidence in rocks of macro and micro fossils but chemical fossils are more elusive and require the availability of appropriate electronic instrumental techniques. Of the reduced carbon in the surface of the planet, living things and dissolved carbon in the oceans constitute a very small amount, whereas the crust of the Earth contains a vast amount of reduced carbon (Table 1). The carbonate, represented by limestones and chalks which are manifestly of biogenic origin, is not

Table 1 *Reduced carbon (graphitic and organic) in the crust of the Earth (after ref. 8)*

		10^{18} *g*
Surface		
Living things (land and water)		0.3
Dissolved C (in oceans)		0.7
Crust		
Volcanics (e.g. basalt), granites and metamorphic rocks		7,000
Sedimentary rocks		
Claystones & shales		8,900
Limestones & dolomites		1,800
Sandstones		1,300
Petroleum		
Non-reservoir		200
Reservoir		1
Coal		15
	Total	19,000

Abundant organic debris is produced in sedimentary rocks of all ages—even back to the Archaean (> 2.5×10^9 years ago). The reduced biogenic carbon in ancient rocks greatly exceeds present-day biospheric carbon in total quantity.

included. We are considering carbon as graphite, diamond and, above all, as enormous quantities of organic debris. Most of this is in the sedimentary rocks, in the form of claystones and shales; these when deeply buried, can give rise to petroleum. Of course, coal represents the most gross form of preservation of organic debris. Note that only a small amount of carbon cycles very quickly at the surface in comparison with the vast amount of buried biogenic carbon which awaits extremely slow recycling over geological time.

Molecular signature of life now

Botryococcus braunii (Figure 1) is a contemporary microscopic unicellular alga that can produce very large quantities of an oil. The algal cells live colonially and form and secrete vast quantities of oil of a very interesting type. Microscopic study of the fossil record shows abundant evidence of a colonial alga that looks very similar to *B. braunii*. This dates back to Permian, but the type fossil comes from Scotland, the Torbane Hill mineral. The chemical evidence is what is really wanted. Does the *Botryococcus*

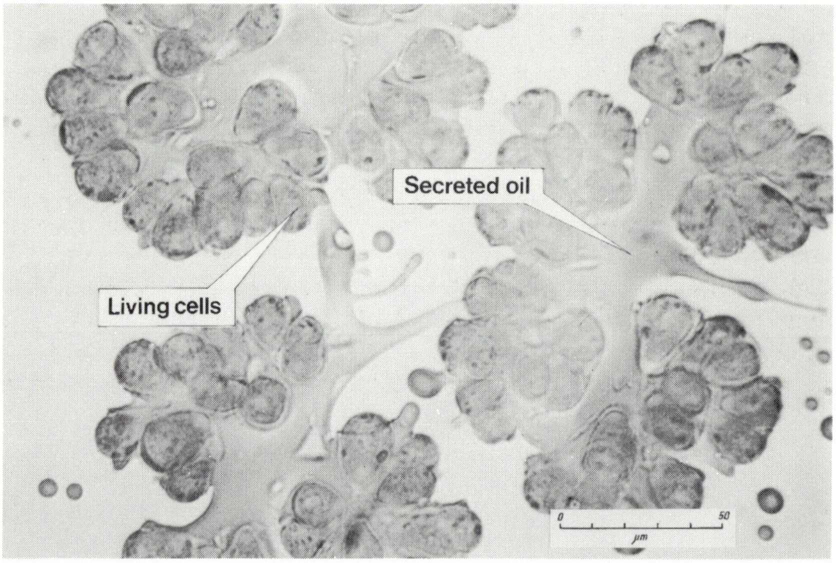

Figure 1 Micrograph of *Botryococcus braunii*, as colonies of cells surrounded by secreted oil. The oil contains unusual hydrocarbons, notably botryococcene and isobotryococcene, which are C_{34} compounds. The reduced fossil counterpart, botryococcane, has been reported in an ancient petroleum. (Courtesy of Dr M. Calvin, University of California, Berkeley, California, USA.)

alga in the living form leave a record of its activities in its chemistry that we could find in ancient rocks?

The question then is which of the analytical procedures should be used for the determination of the molecular record of life; how should the data be interpreted and present-day biosignatures assigned?

Obviously, an organism is best represented by its DNA, then RNA, proteins and so on down the information tree (Figure 2). But unfortunately most of these molecules are very sensitive chemically and will not be well preserved in the geological record. We therefore have to look to the lipids because these are formed of carbon–carbon bonds, in general, which are strong and can withstand the long burial times. Lipids formed by organisms, such as those in the algae discussed above, show great variety from one organism to another but exhibit some similarity in their molecules, so we have a considerable range of parameters available. These are environmentally linked because organisms often use lipids as membrane components which then have to withstand the environment in which the organism exists. Lipids are also used as storage materials, but the most characteristic are those that reflect the environment.

In terms of biomass, microorganisms must dominate the geological record, determining data input and modification of input. A single microorganism contains myriads of carbon–carbon bonds. A lipid molecule is slightly bigger but the cell is so large, relatively that much space is available. There is plenty of scope for the storage of many different lipid molecules in the cell which will, therefore, provide much information (Figure 3).

The molecular components of a cell, much as that of the bacterium *E. coli* consist mainly of water, with about 15% proteins of which there are about 3,000 different kinds estimated for the cell. Much less lipid (about 2% by weight) is found in the cell with about 40 different molecular species. Having found these 40 in an extract, could we say without looking at the rest of the organism, that this is *E. coli*? This is the subject of chemotaxonomy; it is a big challenge and 40 different varieties of molecule are not

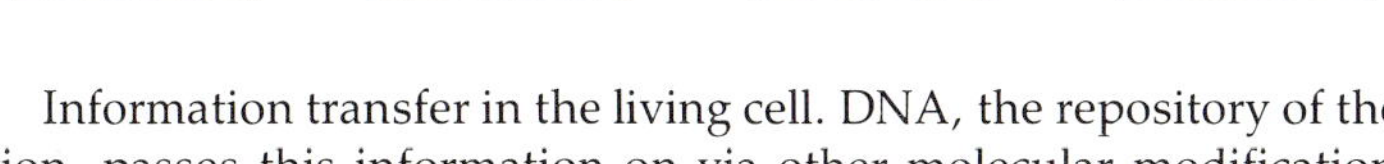

Figure 2 Information transfer in the living cell. DNA, the repository of the genetic information, passes this information on via other molecular modifications in the form of RNA and proteins to generate secondary metabolites, such as lipids. Lipids molecules, unlike the DNA, RNA and protein molecules, are largely structures of carbon–carbon bonds, forming a strong carbon skeleton, not readily broken down by hydrolysis or similar common cleavage reactions.

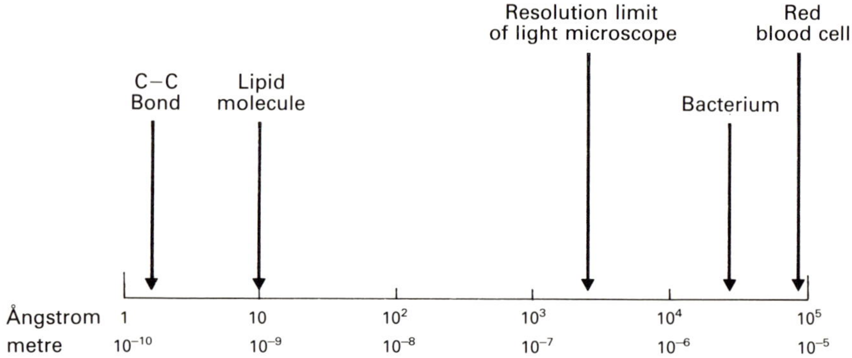

Figure 3 Size in biology. Lipid molecules are very much smaller than a bacterial cell. Hence there is enormous scope for molecular variety, even in the storage lipids of cells.

really enough for this purpose. If we imagine an environment containing a variety of organisms, could we infer the variety present from the molecular composition? (Figure 4).

This approach is not as hopeless as it might first appear. Consider, for example, the number of alkanes possible for each carbon number, as compared with the number observed to occur biologically. There is a high degree of selectivity in the biology which reflects the control of the DNA. So instead of an impossibly large number of different molecules to consider, we encounter a great deal of specificity and order. We are, in fact, observing the end result of information transfer from DNA to lipid synthesis.

Lipids are distributed in various parts of a cell but especially in the cell coat. This is rich in lipids because at this point the cell is in contact with its environment. A barrier or a mechanism for transfer must, therefore, be designed (bio-engineered), and requires special polar molecules such as phosphlipids and glycolipids. The polar and non-polar groups will assemble into membranes where the chains link across the double membrane with the polar heads on the outside and the inside. This piece of biochemistry is important because it links environment and molecular structure. Cells exist in a great variety of environments and, one might suppose that their membranes would show a similar range of diversities in molecular structures which would help if one was examining the lipids to trace the environments that these organisms inhabited.

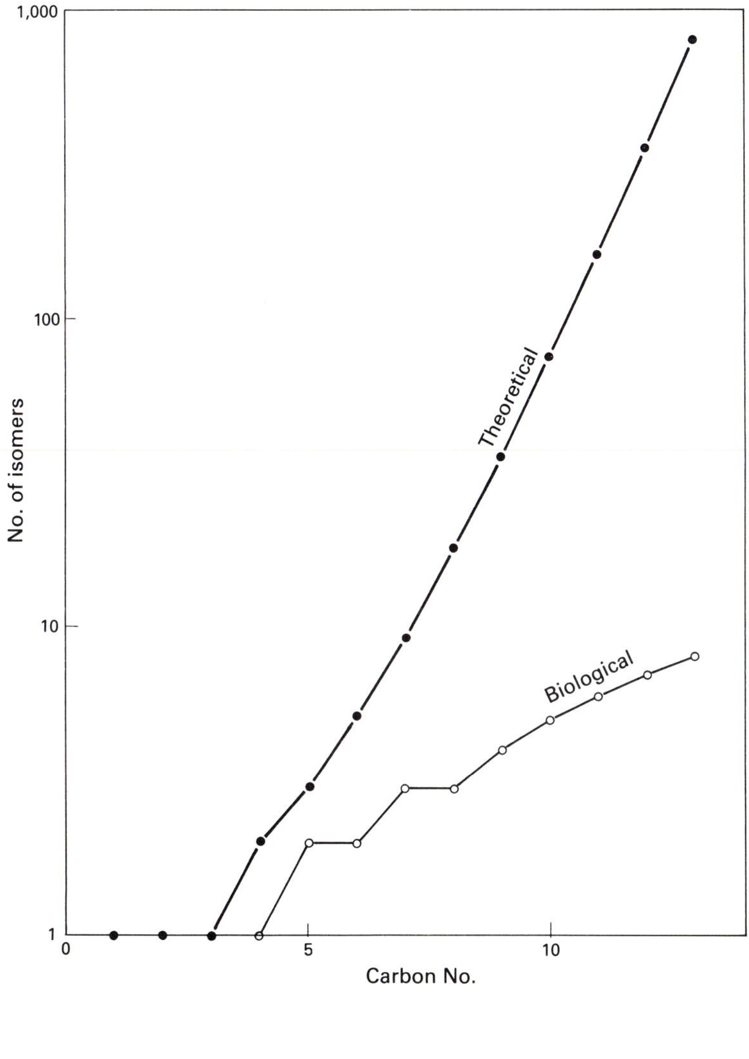

Figure 4 Biological subset of alkanes. The specificity of biosynthetic processes is well illustrated by these plots for the number of isomers (log scale) of a given carbon–number alkane, calculated on a theoretical basis and from direct observation of biological occurrence.

Figure 5 shows temperature, amoxicity and oxicity (in other words Eh), acidity, pressure and salinity plotted on one diagram. Can an organism be represented by a line that joins up the different points where its preferred environment intercepts these axes?

Figure 5 also shows, as a dashed line, the most favourable human environment. But note that halophilic bacteria exist, also organisms that live in virtually no salinity at all, under various pressures and so on. The most controversial area is that of temperature; for example, organisms are found in the deep sea bottom hot springs. At temperatures reportedly well over 200°C, they not only exist but gladly multiply, so how molecularly does the life form manage to cope with that environment?

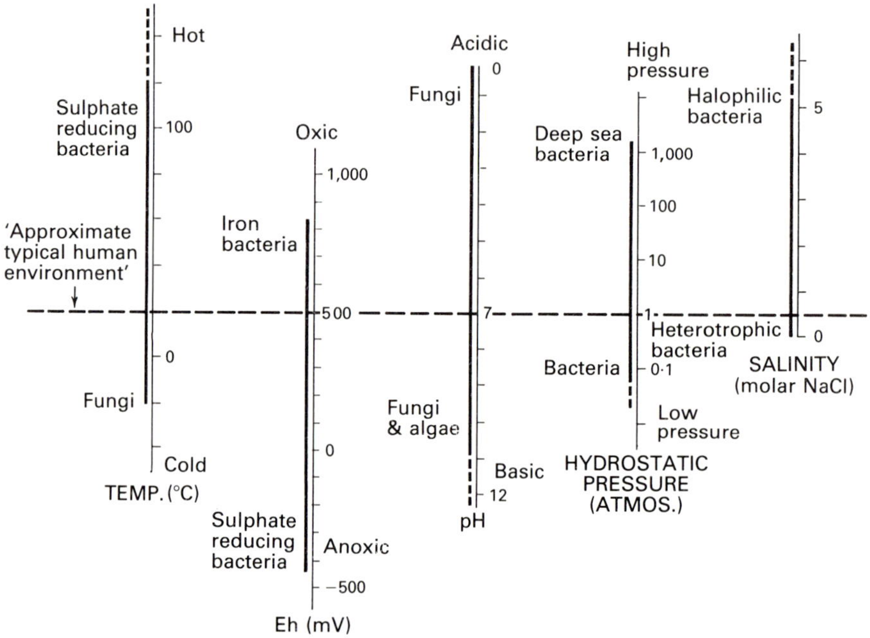

Figure 5 Organisms and environmental extremes—the robustness of life. Various life forms, notably microbial, can exist and even flourish in environments where physical and chemical parameters are in human perception, extreme. The molecular structures and compositions of the biological constituents of organisms show some relationship with their environmental situations, being in part an evolutionary response to them.[5,6]

Determining the molecular signature

We now want to look at these molecular signatures: first at lipids in organisms in various environments, and then in ancient sediments. From there we can consider how to acquire data. The instrument that has really made this work possible, is the computerised gas chromatograph–mass spectrometer (GCMS). The next step is the computerised data base approach for the data which it produces. This should make it possible to use molecular abundancy as a source of information, not only for living things and medical situations but also environments and paleo-environments. It would open the door to thousands of parameters which can be measured more or less simultaneously.

The GCMS consists of a capillary column connected to a mass spectrometer. The mixtures of compounds are separated into the individual molecular species by the capillary column and then enter the mass spectrometer where they can be identified by their mass spectrum, through electron impact, or some other process. Two main parameters are available for the identification of a compound: first the emergence time relative to a standard coming through this column, and second, the mass spectrum which is an array of up to, say, 500 parameters. The process consists of first injection of the sample, second initial processing at the VDU terminal, and third, advanced processing of the elution record of the compounds into the mass spectrometer. Thus a reconstituted ion current or a total ion current of trace is obtained.

To do this process manually a scan is taken and the mass spectrum interpreted using the operator's experience, or, failing that, the machine is instructed to search through the library spectra (say 30,000, most of which are useless!) but eventually a best fit is found. To proceed further, it is possible to look at individual ions and then run through the whole analysis to see compounds eluting in sequence. Thus we can obtain a sort of fingerprint consisting of the relative abundancies of these components. This fingerprinting is very valuable visually but what is really wanted is a list of components with their amounts entered directly into a database (Figure 6).

Interpretation of molecular abundance patterns

Moving forward to molecular assessment, we need to determine molecular abundancies. First, we must ensure that enough of each compound is available for measurement. Then we want to know whether or not the patterns are characteristic of particular organisms. The discriminatory nature of the process will depend on those very patterns. For example, a lipid inner cell wall may well be very different from one organism to another. It is important that there should be some sort of consistency; that

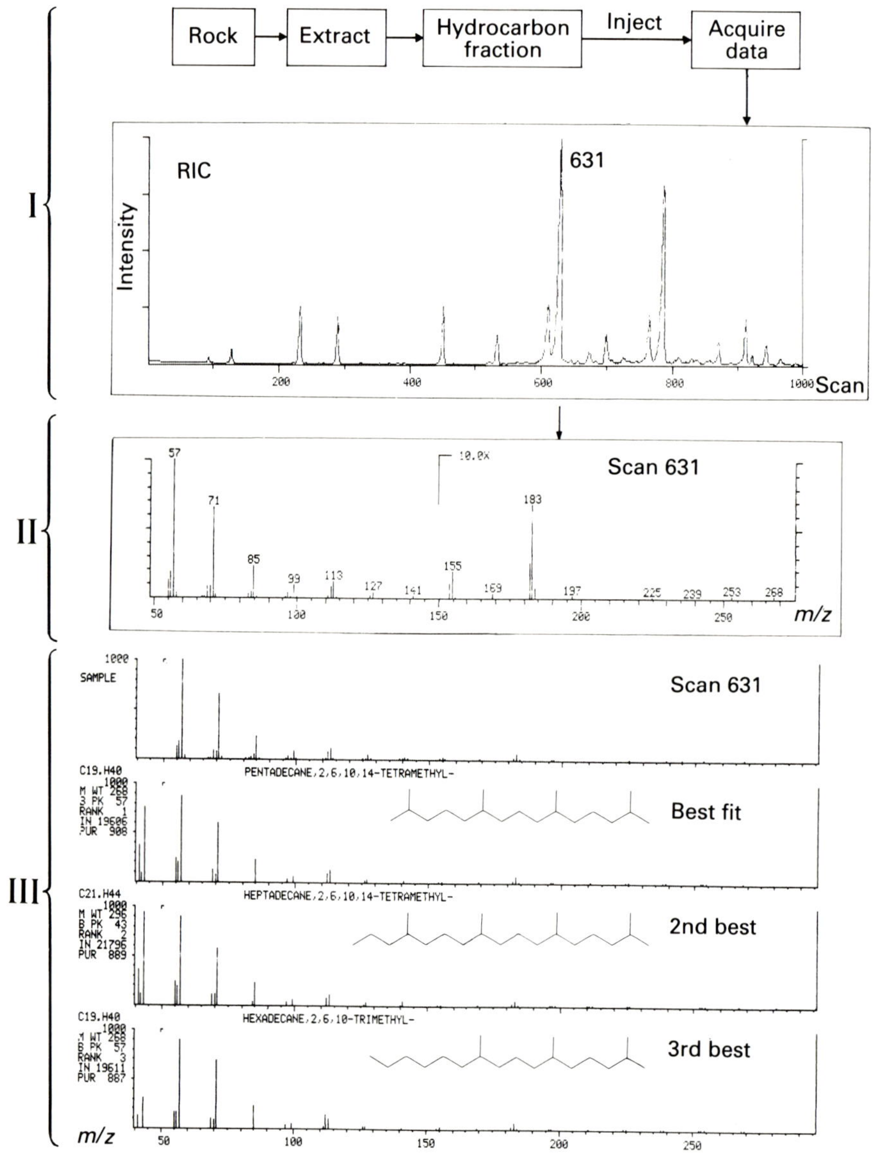
I
Rock
Extract
Hydrocarbon fraction
Inject
Acquire data
RIC
631
Intensity
200
400
600
800
1000
Scan
II
57
71
85
99
113
127
141
155
169
183
197
225
239
253
268
10.0X
Scan 631
50
100
150
200
250
m/z
III
1000
SAMPLE
Scan 631
C19.H40
PENTADECANE,2,6,10,14-TETRAMETHYL-
Best fit
C21.H44
HEPTADECANE,2,6,10,14-TETRAMETHYL-
2nd best
C19.H40
HEXADECANE,2,6,10-TRIMETHYL-
3rd best
m/z
50
100
150
200
250

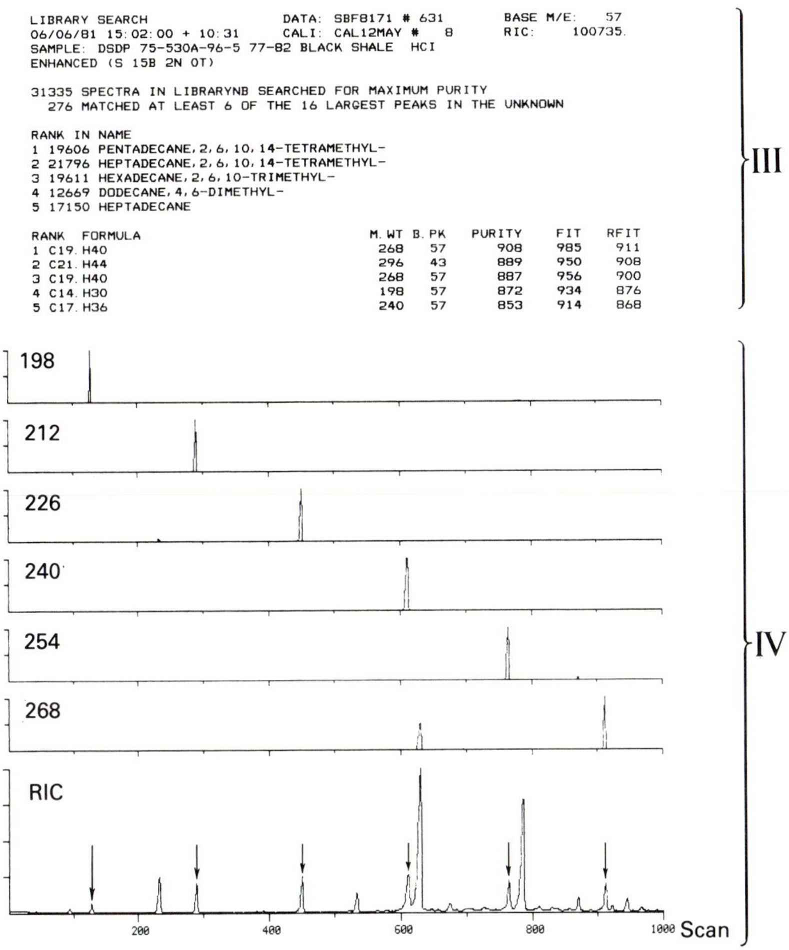

Figure 6 Computerized gas chromatography–mass spectrometry. The example chosen shows the analysis of a hydrocarbon fraction extracted from a black shale sample of Cretaceous age (96 million years old) drilled from the floor of the Atlantic Ocean in the Angola Basin off West Africa. Four stages in the analysis are shown, involving acquisition and data processing with a Finnigan INCOS data system. RIC, reconstituted ion current trace. Scan, individual mass spectrometric scan of about 1 second duration. In stage IV, individual mass chromatograms are shown e.g. 198 = time plot (scan numbers) for *m*/*z* 198 signal.

Table 2 *Molecular assessments. Interpreting the record in the sediments*

Criteria for use of molecular fossils

1. Readily extracted in sufficient quantity to be recognized and quantitated by available methods → *Molecular abundance patterns.*
2. *Molecular abundance patterns,* while characterizations of life in general, should also be sufficiently *discriminatory* for particular species of organism.
3. Palaeo-interpretation requires paradigm of *constancy* of these *molecular abundance patterns* i.e. constancy of biosynthetic pathways and their selective operation in particular species of organism.
4. The *changes* in the *molecular abundance patterns* with diagenesis and then catagenesis (changes with increasing depth of burial) must be understandable in terms of specific reactions.

is, that the same species of organism will always produce the same components, but this is not always the case. Finally comes the very serious problem of change with burial which may in some cases, prove to be an advantage. The problem is that the molecules, when they are buried, are altered in the Earth's crust, changing from something that we know (the biolipid) to something we will have to look for (the geolipid). Furthermore, we shall need to try to make an inference between precursor (biolipid) and product (geolipid).

The advantage arises when those processes of alteration (generally called diagenisis, catagenesis and maturation) are understood and may be used in exploration. True abundance patterns for precursor and product, arrived at by distinguishing signal and noise, should permit a novel molecularly based interpretation of geological materials in terms of events and processes (Figures 7 and 8).

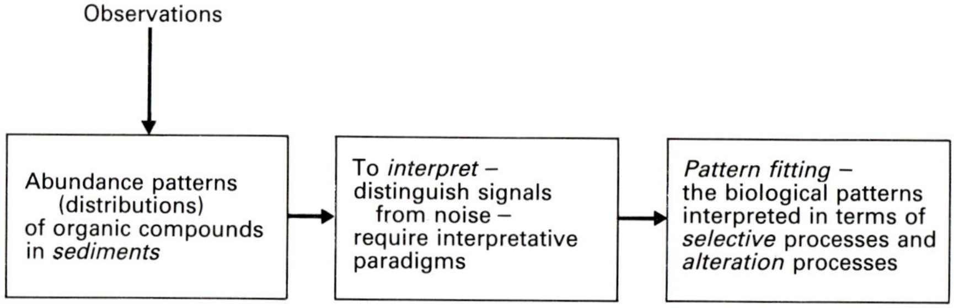

Figure 7 Organic geochemistry and uncertainty: signals/noise.

Figure 8 Molecular assessment of depositional palaeoenvironment. Anoxic bottom conditions are inferred from the lipid record—good preservation of the rain of biological debris reaching the ocean floor. (Based on an original by Timothy Ian Eglinton.)

Catalysis can complicate the picture in terms of organic geochemistry, when we have to guess what we started with, without even knowing which catalysts are present in the Earth's crust. However, some progress has been made and molecular parameters based on the diagenetic conversion of certain lipids to others, are now used by major oil companies in their exploration for petroleum.

We can now consider maturation parameters, using a series of molecular markers. These will change in different parts of the sedimentary column, some very quickly, others take much longer requiring higher temperatures. This produces a series of maximum and minimum thermometers but which are also related to time. These are useful in that, if the molecules being examined have undergone changes within a certain time, then it must lie within a particular zone of diagenesis or maturation of organic matter. This can be very important when trying to explore for oil—which material has given rise to oil or will give rise to oil.

By way of example of these ultra-slow molecular changes, generally

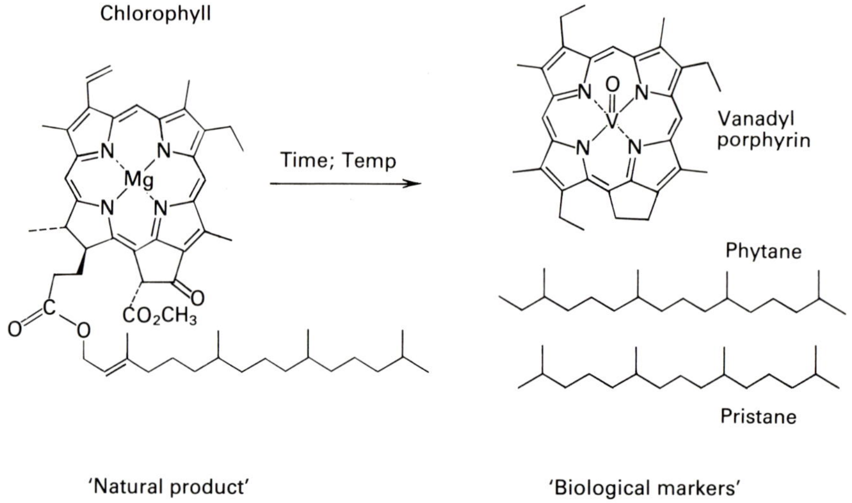

Figure 9 Example of geological fate of biolipids: the green pigment of plants. Chlorophyll *a* is inferred to be the original source of the abundant petroporphyrin pigments (illustrated here is a single vanadyl porphyrin) and two hydrocarbons, pristane and phytane in certain petroleums and ancient shales. The 'natural product' chlorophyll and the 'biological marker' geolipids are assigned a 'precursor–product' relationship.[7]

taking millions of years, take the fate of the molecule chlorophyll when it is buried. We find that the hydrocarbon part (alkyl side chain—phytol which esterifies the chlorin acid) gives rise to hydrocarbons pristane and phytane, while the chlorine part gives rise to reddish purple pigments—petroporphyrins. Burial of a chlorophyll-rich material will, thus, produce two very different types of molecule, which can then be looked for (Figures 9 and 10).

How should all these ideas be applied? We can start with the sediment, do the extractions and record the information, but really basic information is needed from living organisms by way of a separate data base, and then we need to look at modern environments, extract the bottom sediments and obtain similar sets of information which will help to interpret the palaeo record. We must extend our activities into living systems and into various environments (Figure 11).

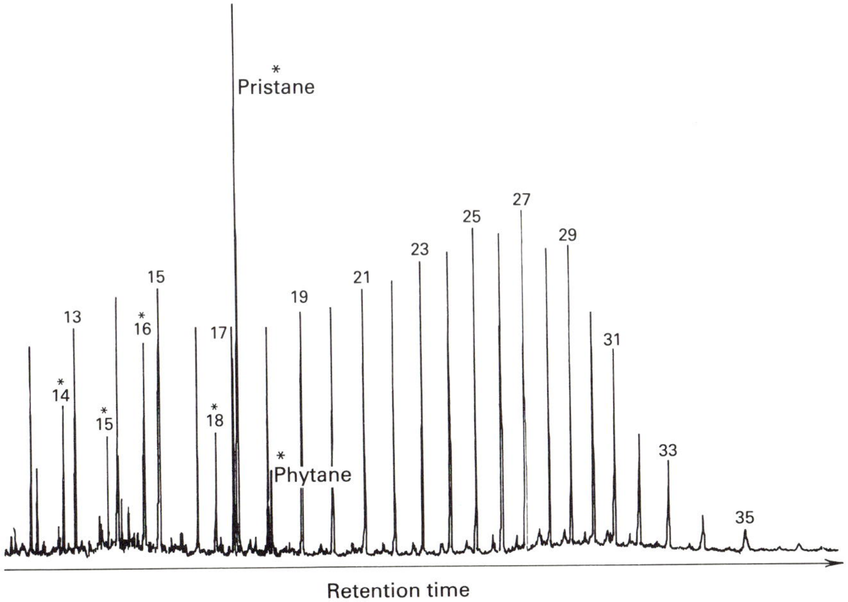

Figure 10 Analysis of the alkane fraction of a young crude petroleum by gas chromatography. Note the abundance of pristane, phytane and other isoprenoid hydrocarbons (*) believed to be derived, in part at least, from chlorophyll.[7]

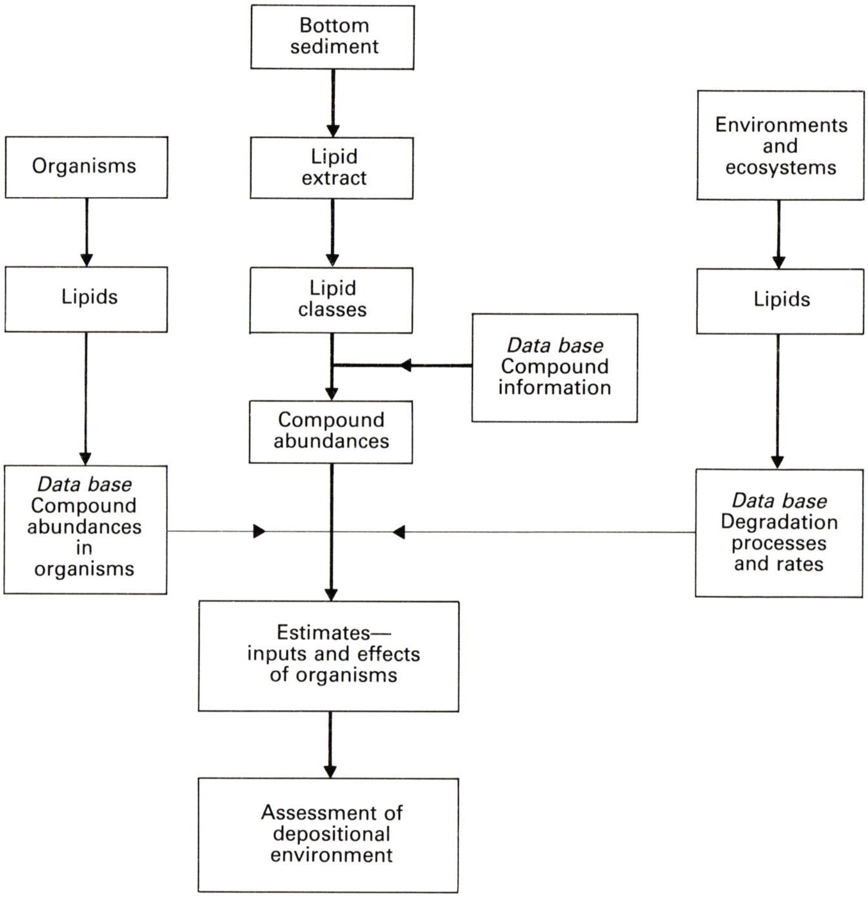

Figure 11 Molecular assessment of aquatic sediments. Proper interpretation of the molecular record of lipid inputs to the ancient sediments—the geological record—will require complicated manipulation of the geological database in parallel with databases of compound abundances for living organisms and aquatic ecosystems and environments.

Data matrices

Figure 12 shows a series of straight chain hydrocarbons as an example of this approach. In our assessments we can use C_{10} to C_{40} hydrocarbons as one simple set of biological marker molecules. If we now functionalise these with hydroxyl or carboxyl groups or some other functional group, we create further sets of molecular parameters (Figure 13). In the higher plant data shown, the *n*-alkanes display a marked odd over even predominance in the carbon number range C_{23} – C_{35}, whereas the *n*-alkanoic acids show a dominance of even carbon number homologues. The other series of compounds also show interesting relationships. Such distributions of homologues are characteristic of higher plants growing in a terrestrial environment. The aquatic unicellular algae present a very different situation. Such planktonic organisms generally contain shorter chain compounds, except for the unusual *n*-alkenones. A preliminary, simplistic approach to the assessment of biolipid contributions to a Recent sediment can now be made by examining the distribution of the six series of

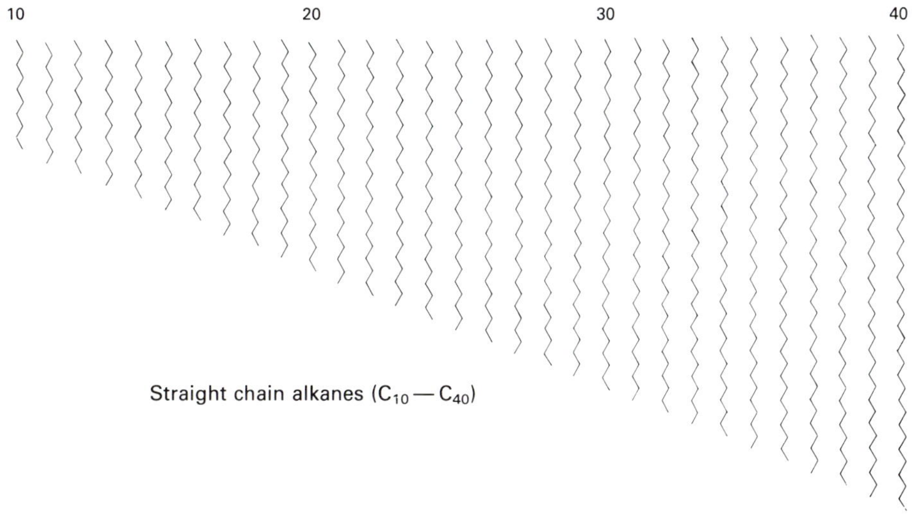

Figure 12 A pattern of molecules. The full data set of the straight-chain saturated hydrocarbons, the n-alkanes, ranging in chain length from C_{10}–C_{40} is drawn here. This set of hydrocarbons is readily analysed for, providing a set of 30 molecular parameters, the relative or absolute abundances of which can be entered into a database for statistical manipulation, along with other molecular data sets and items of information, for a sample under study.

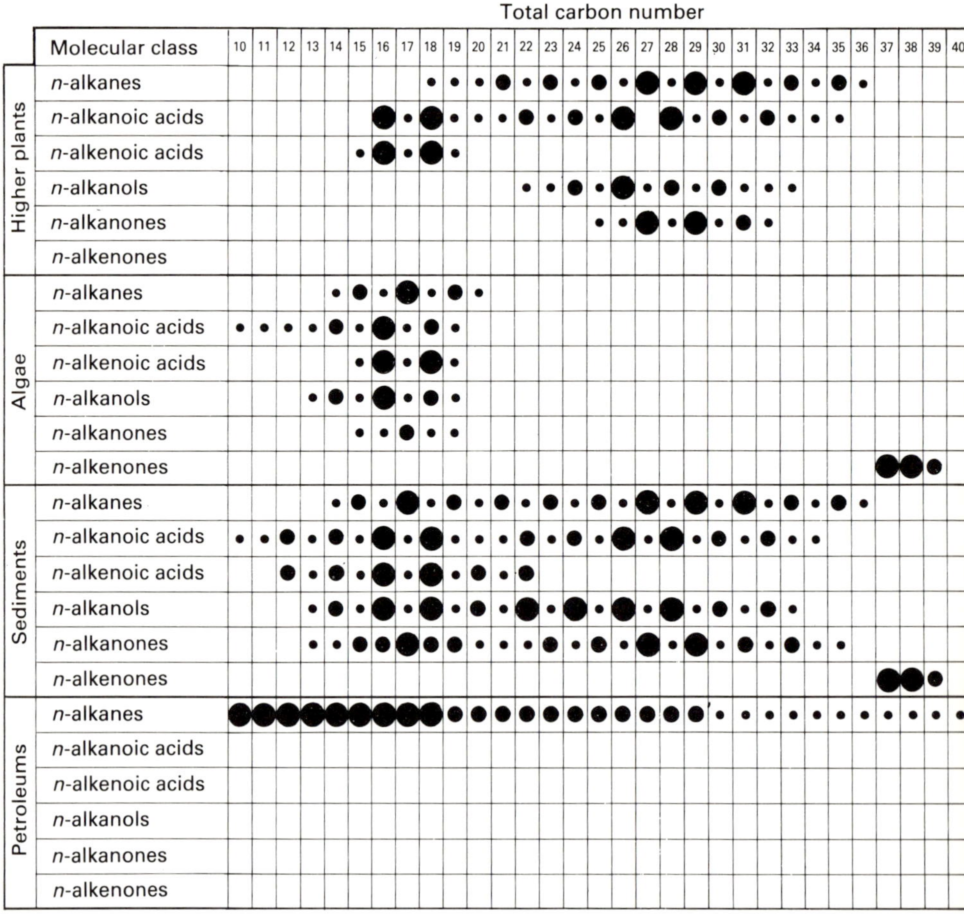

Figure 13 Multimolecular parameters: carbon number/abundance matrix for straight chain compounds. This data set refers to the relative abundances of individual homologues of the $C_{30} - C_{40}$ *n*-alkanes (*cf.* Fig. 12) and the corresponding straight chain alkanoic and alkenoic acids, alkanols, alkanones and alkenones. The circles provide a very approximate idea of quantities, • small, ● medium and ⬤ large, on a tenfold scale. The data were compiled by Dr S. C. Brassell, using literature data as an approximate basis. An abundance matrix of this type provides a ready means of visual comparison between samples. However, with more extensive data sets for additional molecular parameters, such as acyclic and cyclic isoprenoids, multiple visual comparisons become impracticable. Computerised, statistical assessment then becomes highly desirable, if not essential, for effective evaluation.

homologues in the lipid extract. Petroleums, on the other hand, which do not contain appreciable amounts of the functionalised molecules, show some indication of input from the *n*-alkane distributions, even though the odd–even predominance has largely disappeared. Many parameters must be considered simultaneously to assess the situation. A database approach has to be developed, as proper assessment of the huge amount of data cannot be done visually. Database methods are well established but major problems remain of entering, storing and assessing data, prior to interpretation. The fast-developing subject of chemometrics, where chemical abundances are treated numerically, provides one approach. Two main chemometric procedures are especially promising for the manipulation of multimolecular data, *viz.* pattern recognition, through Principal Component Analysis and Cluster Analysis, and transform techniques, for example Fourier transforms.

Examples of multimolecular analysis: climatic signals

In recent work at Bristol we have conducted lipid analyses on cores of sediment drilled from the coean floor. Some of these cores exhibit fine laminations corresponding to layers of diatomaceous ooze, organic matter, clay minerals, calcium carbonate and so on. But, can the molecular abundance data be used to extract palaeoclimatic information from such cores? Palaeoclimate assessment has already been achieved rather successfully using stable isotopes, such as oxygen-18 (see, for example, refs 1–3).

Fourier transform techniques, when applied to the downhole record, reveal recurring cycles—periodicities in the data—for 41,000, 23,000 and 19,000 years, very probably related to astronomical forcing functions such as the obliquity of the ecliptic for the first, and precession of the equinoxes for the last two.

Obviously, if molecular data are also going to be utilised in a similar fashion, then this is the sort of technique which should be applied to them. The proposed link between the molecular sedimentary record and the palaeoclimatic record is made as follows: from the relationship between environment and membrane structure, we know that organisms counter changes in temperature by changing their lipid composition.[4] Hence, analyses of biolipids can be used to infer water temperatures. Water temperatures, in turn, are related to the overall global climate. Finally, sedimentary lipids (geolipids) are a partially preserved record of the biolipids and hence, by inference, palaeoclimate.

Simon Brassell and Ian Marlowe at Bristol have looked at a sediment core taken off Africa and have plotted a molecular unsaturation parameter, U°, measured for compounds of phytoplankton origin, against section depth. They found that the degree of unsaturation oscillates in a periodic fashion over the past million years.

The paradigm we have adopted involves sequential causal relationships between astronomical forcing functions, climate, sea surface temperatures, lipid composition of the phytoplankton that grow in these waters, and, finally, the multimolecular data for certain of the long chain phytoplankton lipids found in the bottom sediments. The rationale for this study, presently in progress in collaboration with Sarnthein and his colleagues at Kiel, is that the composition of the lipids biosynthesised by phytoplankton in the zone of primary productivity of the oceans will reflect the water temperature prevalent in that water mass. This inferred relationship between water temperature and lipid composition during growth is based on a number of direct observations and experimental studies. In brief, certain, very long, straight chain C_{37} – C_{39} ketones biosynthesised by the green alga, *Emiliania huxleyi*, and other marine unicellular algae, are found to have a higher U° index, with lower water temperatures and *vice versa*. Our working hypothesis is that variations in °C, found for the sedimentary lipids, can be assessed as variations in palaeotemperatures for such sites. We are currently seeking to test this hypothesis over the last 1 million years against the accepted ^{18}O index.

Example—oil generation—basinal

We now consider the generation of petroleum by deep burial of organic debris. This classic generation of oil takes place in sedimentary basins at considerable depths and is a very slow process. Its generation may be predicted by measuring the type, quantity and maturity of the organic matter which is present in the prospective source rocks. Inter-sample comparisons and correlations are also very important; which oil is related to which other oil? Which oil is related to which source rock? These are the critical questions posed by exploration geologists when looking for oil.

In the North Sea basin, for example, the prospective source rocks, which are Jurassic in age, contain abundant insoluble fossil organic matter or 'kerogen'. Examination under the microscope suggests that this amorphous material is mostly of algal origin, but fortunately molecular techniques give more definitive information, allowing estimates of the relative inputs of terrestrial plant debris, phytoplankton (algal), and microbial debris. This potential source rock, rich in organic debris, was progressively buried deeper and deeper as fresh sediment was added in the North Sea above, its temperature rising steadily higher, until eventually this slow cooking gave oil. The soil then migrated out of the source rock into various faults and so reached sands and other permeable and porous formations where the oil remained in a 'reservoir'. This very slow process of oil generation can be rationalized in terms of the combined effects of time and temperature (Fig.14 and Display).

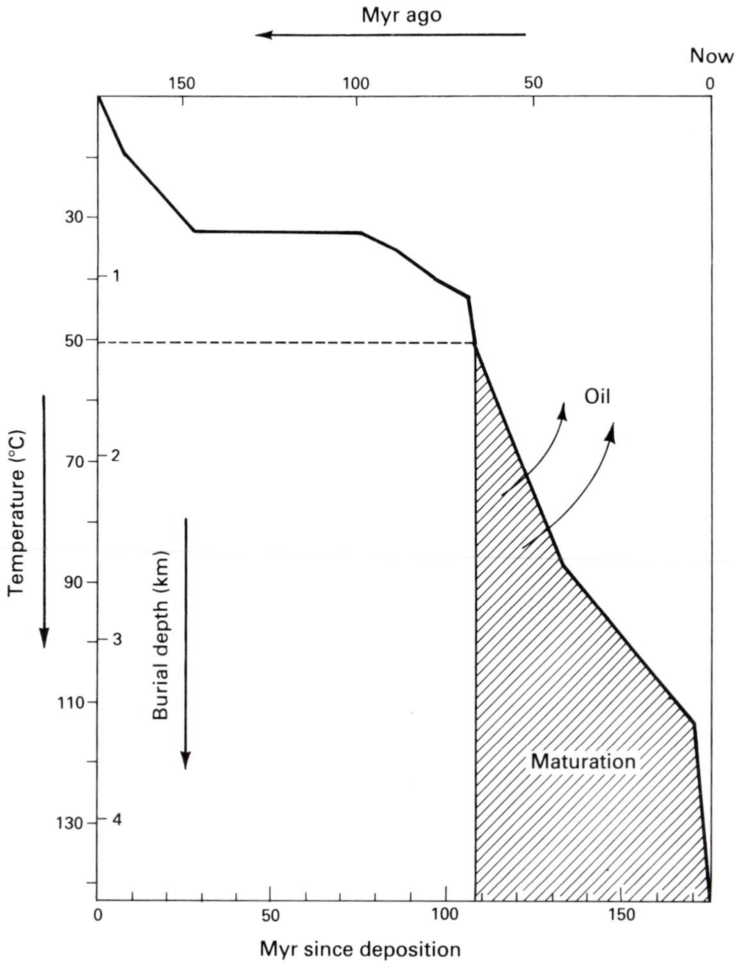

Figure 14 Subsidence and maturation history of Jurassic source rocks in North Sea graben. The source rocks for some North Sea oil reserves are widely believed to be Upper Jurassic, Kimmeridgian or slightly older in age (about 155 Myr). These sediments when deposited, were laid down in an off-shore basin and contained both abundant higher plant material of terrestrial origin and amorphous organic debris of probable phytoplanktonic origin. They have achieved their deepest burial in the down-faulted Central Graben, which runs approximately N-S between the UK and Norway, where the geothermal gradient and heat flux is such that they are now experiencing temperatures of about 140°C beneath about 4 km of overlying sediments. The subsidence history is believed to have followed approximately the time profile shown, with oil formation and migration beginning some 50 Myr, when maturation of the organic matter had advanced sufficiently. This scenario is matched by the laboratory assessment of the molecular nature and maturity of the organic matter present in source rock samples recovered from various burial depths drilled beneath the North Sea.

"RECIPE" FOR NORTH SEA OIL

987.-NORTH SEA OIL

(Fr.—L'Huile de La Mer du Nord)

Ingredients.—10,000,000,000,000,000 Kg of fine-grained, Kimmeridgian clay, rich in decaying microscopic marine algae (5% total organic carbon).

Mode.—Bury sediment extremely slowly, using more sediment, to 3 Km deep beneath sea floor. Meanwhile warm steadily to 150°C, allowing the pressure to rise to 500 atmospheres (50 MegaPascals). Keep lid on tightly or oil will escape. Serve through a pipe 4 Km long.

Time.—150,000,000 years. The mature sweet oil begins to gather after 50,000,000 years.

Average cost.—ingredients and cooking alone, 0d. After production, $30 or more a barrel.

Sufficient for the whole population (1984).

Seasonable from late 1970's to ca. 2000 A.D.

Note.—Only a portion of the oil can be got out of the ground.

Apologies to:– Mrs. Isabella Beeton.

"THE BOOK

of

HOUSEHOLD MANAGEMENT"

Ward, Lock and Co. London 1861.

N.B. A rival, laboratory-based, 'recipe' is given by Saxby & Riley in ref. 15. Sediments are 'cooked' for 6 years, oven temperatures being raised from 100–400°C at 1°C per week.

The molecular distribution patterns of key biological marker compounds, notably steroid and triterpenoid alkanes and mono- and triaromatic steroid hydrocarbons, have been used in assessing both the origin of the major oil accumulations in the Northern North Sea area and the burial history—rates of burial *vs.* uplift—of potential source rocks for these oils (see refs 9, 10,11). The measurements were arrived at by computerised GC-MS analysis of soluble extracts of the sediments and oils; detailed study allows determination of "molecular maturation" indices—the extent of configurational isomerisation of, for example, steroid alkanes and of aromatisation of aromatic steroid hydrocarbons. Such studies are now being widely conducted by the major oil companies and are becoming accepted as a significant part of the exploration approach, especially valuable in arriving at improved estimates of oil reserves in basins worldwide.

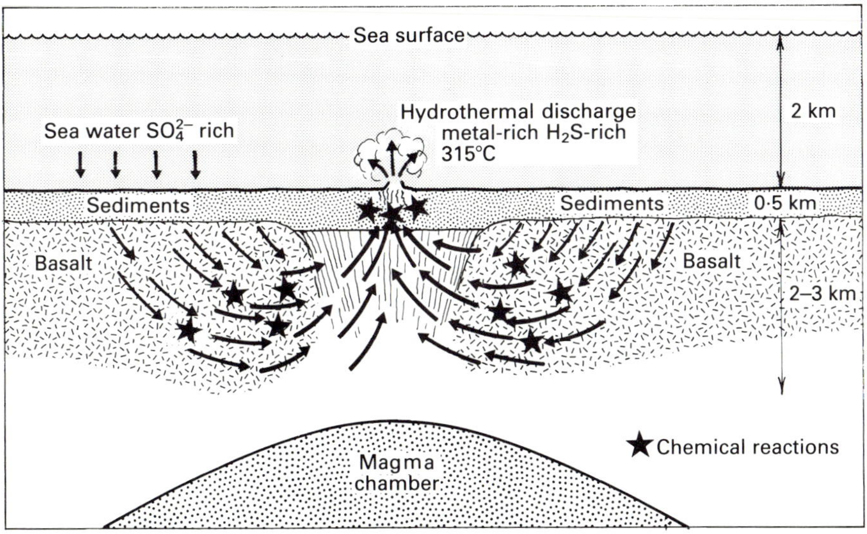

Figure 15 Hydrothermal circulation in the Guaymas Basin of the Gulf of California. The Guaymas basin is an actively-rifting oceanic basin. The young basalt floor (Quaternary age, < 1 Myr) as it forms at the Spreading Centre is being covered rapidly with a blanket of sediment. The high heat flow from the deep seated magma brings about a hydrothermal circulation in which seawater flows down into the basalt and then, when superheated, flows up through the unconsolidated diatomaceous muds, pyrolysing the organic matter and entraining the 'oil' in the hydrothermal discharge at the vents and hydrothermal mounds (Fig. 16).

Example—Oil generation—hydrothermal

Oil can also be generated hydrothermally. This 'instant' oil can be found, for example, in the central Gulf of California, where its generation has taken place rapidly at shallow depths.[12,13] It is a rifting zone which was drilled through about 2 km of water (as Leg 64 of the Deep Sea Drilling Project), that has produced some extraordinary results. The site description is as follows: an enormous influx of fresh water and sediment into the Gulf from the Colorado River has taken place throughout the Recent geological past. The high sedimentation rate, added to by the high productivity—mainly diatoms (fatty, unicellular algae) has produced a thick layer

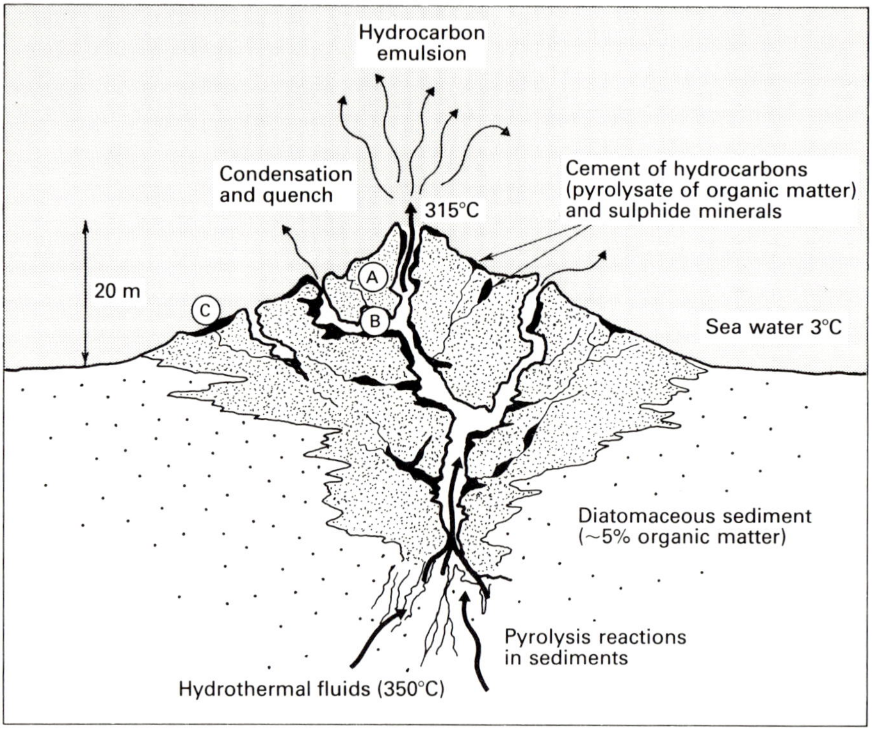

Figure 16 Sub-sea hydrothermal mound. This type of mound occurs at DSDP Site 477 in the Gulf of California (Fig. 15) and has been photographed and studied with the DS RV 'Alvin' of the National Science Foundation. Active mounds continue to grow, as minerals are deposited from the metal sulphide rich super heated waters emerging from the vent. Organic pyrolysates derived from the diatomaceous, organic-rich sediments, also stream from the vent as oil emulsions and gradually form deposits in the mound and outside, where they have been sampled by careful dredging and other means and analysed (Fig. 18). (After refs 12, 13.)

of organic-rich bottom sediment. The Gulf is a high heat flow region with the consequence that seawater, which percolates through the very young sediments down into the basalts beneath, is then heated to high temperatures. A hydrothermal return flow upwards through the young sediments is generated, which carries dissolved metal salts and hydrogen sulphide. Organic matter in these diatomaceous sediments is effectively 'pressure-cooked' very quickly and hydrothermally-generated oil emerges from mineral-rich mounds on the floor of the Gulf, as emulsions (Figures 15 and 16). In such a hydrothermal mound, the minerals deposited include heavy metal sulphides and solid hydrocarbons of a variety of compositions. These 'bitumens' can resemble crude oil but the various fractionation processes in the vent give rise to a wide range of hydrocarbon deposits (Figure 18). We knew nothing of such phenomena until their discovery a few years ago, but a geological parallel has been reported very recently by Adrian Boyce and his colleagues in Glasgow.[14]

The fossil specimens, which exhibit microvents, are found in ore bodies in Ireland from the Carboniferous Period, over 300 Myr. The model that Boyce *et al.* have for their formations calls for hydrothermal action. These ores do indeed contain bitumens which have probably been deposited in similar fashion to that occurring today in the Gulf of California (Figure 17).

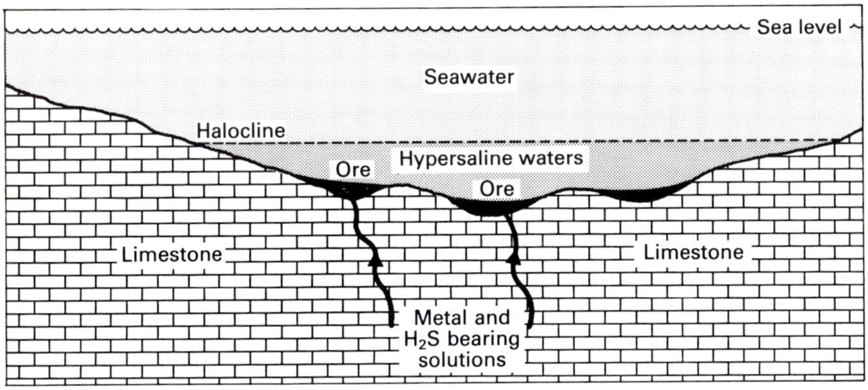

Figure 17 Proposed model of sulphide ore formation in the Carboniferous (Silvermines, Ireland). (after ref. 14). These mineral deposits show chimneys and vents on a micro scale (mm – cm) which, apart from their smaller size, show many similarities with those found in contemporary submarine volcanic vents. Bituminous deposits co-occur with the sulphide minerals, as in the Gulf of California vents and mounds (Figs 16 and 18).

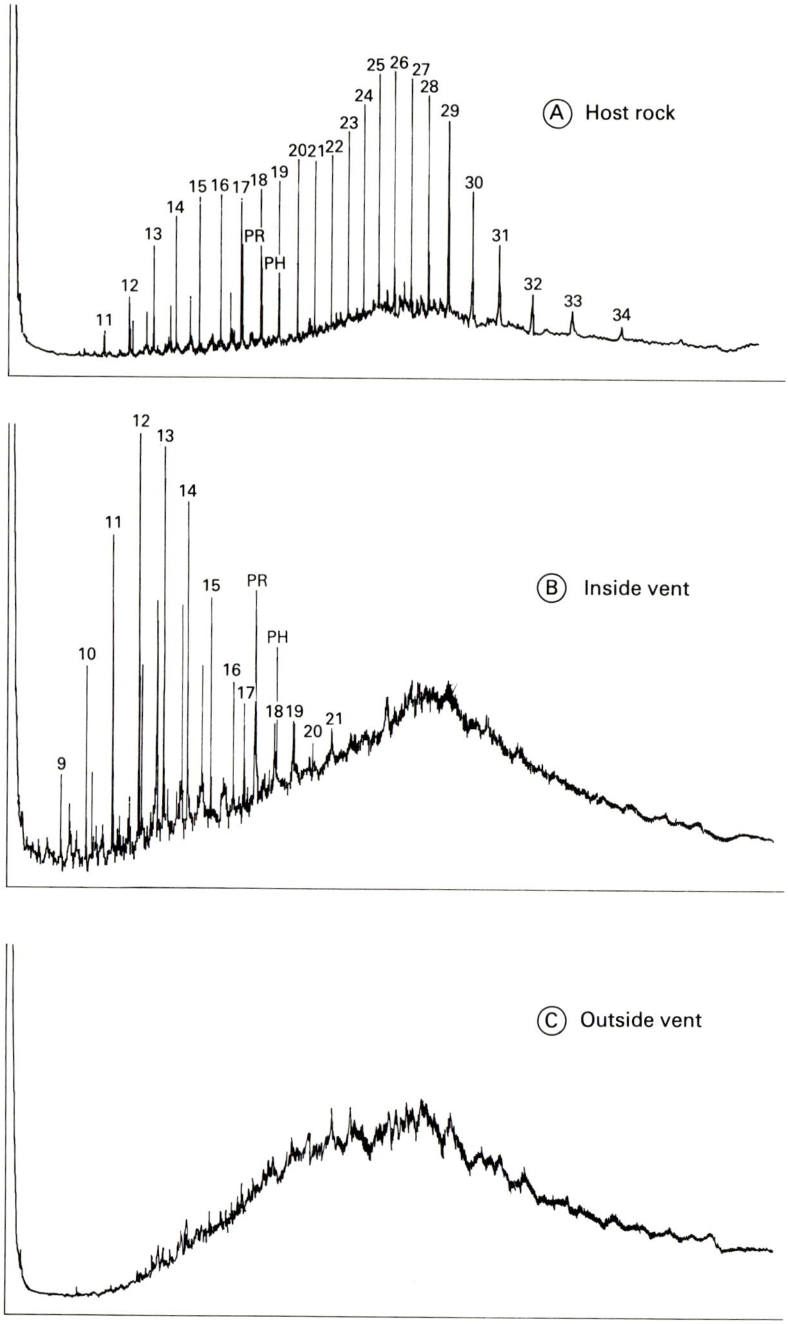
25 26 27
28
24
29
23
20 21 22
19
15 16 17 18
14
13
PR
PH
12
11
30
31
32
33
34
(A) Host rock
12
13
14
11
15
PR
PH
10
16
17
18 19
20 21
9
(B) Inside vent
(C) Outside vent

Conclusions

Living organisms are complex molecular machines. All their attributes, such as reproduction, growth, movement and communication, are molecularly controlled through the action of DNA, RNA, proteins and the other vital macromolecular biopolymers. The single cells of 'simple' microorganisms also contain many thousands of different carbon compounds of medium size—metabolites, membrane components, building blocks, etc.—the presence of which in differing proportions provides a 'molecular fingerprint' of the organism and its activities. These tell-tale molecular abundance patterns are far from thermodynamic equilibrium and, even when taken in isolation, constitute clear evidence of biochemical activity. They even provide some indication of the presence of specific organisms. Thus, some compounds and groups of compounds possess structural and

---

Figure 18 Gas chromatograms for hydrocarbon fractions isolated from bitumens of a hydrothermal mound in the guaymas basin (after Simoneit refs 12, 13). The Hydrocarbon fractions are from:–

A) The rocky matrix (host rock) of the mound—representing the mineral mass deposited by the hydrothermal fluids in contact with the cold bottom waters. Occluded bitumen cementing the minerals shows only slight evidence of biodegradation. The long chain *n*-alkanes (*cf.* Figs 12, 13) indicated by the numbers 11 to 34, show a distribution pattern somewhat similar to that of a crude petroleum (Fig. 10). These *n*-alkanes and the isoprenoids (PR = pristane and PH = phytane, *cf.* Fig. 9) derive from hydrothermal action on the organic matter in sediments beneath the mound.

B) Bitumens deposited inside one of the vents. Here there is evidence of marked fractionation—the large 'hump' corresponds to an unresolved mixture of hydrocarbons; most of the longer chain *n*-alkanes have not been transported to this site.

C) Bitumen deposited outside the vent, away from the present flux of hot waters. Fractionation processes, but especially microbial degradation, have removed almost all of the *n*-alkanes and the acyclic isoprenoid hydrocarbons, leaving the unresolved 'hump' of resistant hydrocarbons.

Computerised GC-MS analyses reveal the above molecular information and much more. Analyses of ancient bitumens in ore deposits such as those of Silvermines, Ireland (Fig. 17) will, it is hoped, reveal similar patterns of molecular abundances matching the proposed similarity in hydrothermal mineral emplacement, now and in the geological past.

stereochemical features characteristic of, say, terrigenous higher plants, unicellular algae such as diatoms and dinoflagellates or bacteria. These compounds constitute specific 'biological markers'.

In recent years, a single powerful technique, that of computerized gas chromatography and mass spectrometry, has made it possible to analyse these complex mixtures. The fat-soluble lipids, whose molecules generally contain 10–50 carbon atoms linked in straight and branched chains, rings and combinations of these, are specially suited to this chemotaxonomic approach for the recognition of organisms. Furthermore, analyses can be fully computerised, molecular abundances being entered directly into databases to await subsequent statistical evaluation. The ability to handle these large data sets interactively would have major consequences for research in biochemistry, biomedicine, environmental chemistry and geochemistry.

These molecular fingerprints persist after the death of the organisms, sometimes showing through in the lipids of organisms higher up in the food chain. Moreover, under the right burial conditions, they may be preserved in sediments as assemblages of 'molecular fossils'.[7,10,11,16-18] This molecular record is readable as the relative abundances of individual compounds, reflecting both the contributions of organisms inhabiting the environment of deposition and subsequent biological or chemical modification during burial. In newly-deposited muds, the record is rich and variable and includes many functionalised compounds derived directly from living organisms. Analysis of older, consolidated sediments shows that some carbon compounds can and do maintain a high degree of structural integrity and sterochemical specificity over hundreds of millions of years. However, when the buried organic matter has suffered raised temperatures by deep burial or contact with igneous intrusions, thermal cracking can give rise to petroleum and natural gas. Even truly ancient rocks of Archaean age—more than 2.5 billion years in age—have been shown to contain molecular evidence of early life.

Acknowledgements

First, my warmest thanks to Leslie Banks of IBM who coaxed me through the preparation of this lecture, helping me in the selection of the material and in the design and presentation of the figures and diagrams. His humourous and questing mind has made its mark on talk and author.

Several close colleagues also helped greatly, notably Paul Farrimond, B.Sc., with the preparation of the figures, and Ian Marlowe, B.Sc. and Drs. Simon Brassell, Richard Brereton, Adrian Boyce and Bernd Simoneit in the provision of key items of information, and Dr. Stan Evans of Finnigan MAT in the display of C-GC-MS equipment.

References

1. Backman, J., Shackleton, N.J., 'Quantitative biochronology of Pliocene and Early Pleistocene calcareous nannofossils from the Atlantic, Indian and Pacific Oceans', *Mar. Micropaleont.*, *8*, 141–170 (1983/84).
2. Shackleton, N., and Boersma, A., 'The Climate of the Eocene Ocean', *J. Geol. Soc.*, *138*, 153–157 (1981).
3. Sarnthein, M., Erlenkeuser, H., von Grafenstein, R., and Schröder, C., 'Stable isotope stratigraphy for the last 750,000 years: Meteor core 13519 from the Eastern Equatorial Atlantic', *Meteor Forsch. Ergebnisse, C38*, 9–24 (1984).
4. Kates, M., Pugh, E.L., and Ferrante, G., *Regulation of Membrane Fluidity by Lipid Desaturases', (1984).*
5. Brock, T.D., *Biology of micro-organisms,* 3rd Edition, Prentice Hall (1979).
6. Vallentyne, J.D., 'Environmental biophysics and microbial ubiquity', *Ann. N.Y. Acad. Sci.*, *108*, (1963).
7. Evershed, R.P., Barwise, A.J.G., Eglinton, G., and Maxwell, J.R., 'Red pigments in petroleum exploration, *Interdisciplinary Science Reviews* (submitted).
8. Hunt, J.M., *Petroleum Geochemistry and Geology*, W.H. Freeman and Co., San Francisco (1979).
9. Mackenzie, A.S., and Maxwell, J.R., 'Biological marker and isotope studies of North Sea crude oils and sediments', in *Proc. 11th World Petroleum Congress*, Vol. 2, pp.45–56, John Wiley Chichester (1984).
10. Mackenzie, A.S., Brassell, S.C., Eglinton, G. and Maxwell, J.R., 'Chemical fossils: the geological fate of steroids', *Science*, *217*, 491–504 (1982).
11. Curtis, C.D., Eglinton, G., McKenzie, D., and Murchison, D.G. (Editors) *Geochemistry of Buried Sediments*, *Phil. Trans. R. Soc. A.* (1985).
12. Simoneit, B.R.T., *Effects of hydrothermal activity on sedimentary organic matter: Guaymas Basin, Gulf of California—petroleum genesis and protokerogen degradation* (1983).
13. Simoneit, B.R.T., 'Organic matter maturation and petroleum genesis: geothermal versus hydrothermal', in Eaton, G. *et al.* (Editors), *The Role of Heat in the Development of Energy and Mineral Resources in the Northern Basin and Range Province*, Geothermal Resources Council, Davis, (1984).
14. Boyce, A.J., Coleman, M.L., and Russell, M.J., 'Formation of fossil hydrothermal chimneys and mounds from Silvermines, Ireland', *Nature*, *306*, 545–550 (1983).
15. Saxby, J.D., and Riley, K.W., 'Petroleum generatin by laboratory-scale pyrolysis over six years simulating conditions in a subsiding basin', *Nature*, *308*, 177–179 (1984).
16. Eglinton, G., Hajibrahim, S.K., Maxwell, J.R., Quirke, J.M.E., Shaw, G.J., Volkman, J.K., and Wardroper, A.M.K., 'Lipids of aquatic sediments, Recent and ancient', *Phil. Trans. R. Soc. Lond. A.*, *293*, 69–91 (1979).
17. Brassell, S.C., Eglinton, G., and Maxwell, J.R., 'The geochemistry of terpenoids and steroids', *Biochem. Soc. Trans. 11*, 575–586 (1983).
18. Ourisson, G., Albrecht, P., and Rohmer, M., 'The microbial origin of fossil fuels', *Scient. Am.*, 251(2), 34–41 (1984).

Contributors to
Science and Uncertainty
in timetable order

Professor Eric Laithwaite,
A scientist is often wrong but never in doubt;

Dr. Graham Cairns-Smith,
On the rocks: the vital cocktail?;

Professor Geoffrey Eglinton,
The alchemy of life: probing the molecular record;

Professor Ray Lyttleton,
Journey to the centre of uncertainty;

Dr. Richard Voss,
Random fractal forgeries: from mountains to music;

Professor Bob Churchhouse,
Uncertainty in mathematics and computing;

Professor David Kendall,
Chance or design?;

Dr. Geoff Robinson,
Perception from pictures;

Professor Josef Raviv,
Modelling for uncertainty reduction;

Jim Baldwin,
Fuzzy reasoning: the need for blunter tools;

Professor Henri Nussbaumer,
Digital signal processing and computational complexity;

Dr. Peter Burrough,
Fakes, facsimiles, and facts: fractal models of geophysical phenomena;

Professor Norman Dixon,
Consciousness: the illusion of certainty;

Dr. Richard Garwin,
Man, technology, and uncertainty.